139 Topics in Current Chemistry

Organic Geo- and Cosmochemistry

With Contributions by
E. Herbst, F. Mullie, T. Nakashima,
H. D. Pflug, J. Reisse, G. Winnewisser

With 53 Figures and 14 Tables

Springer-Verlag Berlin Heidelberg GmbH

This series presents critical reviews of the present position and future trends in modern chemical research. It is addressed to all research and industrial chemists who wish to keep abreast of advances in their subject.

As a rule, contributions are specially commissioned. The editors and publishers will, however, always be pleased to receive suggestions and supplementary information. Papers are accepted for "Topics in Current Chemistry" in English.

ISBN 978-3-662-15174-7   ISBN 978-3-540-47197-4 (eBook)
DOI 10.1007/978-3-540-47197-4

Library of Congress Cataloging-in-Publication Data. Organic geo- and cosmochemistry. (Topics in current chemistry ; 139
1. Organic geochemistry. 2. Cosmochemistry. I. Herbst, Eric. II. Series. QD1.F58 vol. 139 [QE516.5] 540 s [551.9] 86-24771
ISBN 978-3-662-15174-7

Typesetting and Offsetprinting: Th. Müntzer, GDR;
Bookbinding: Lüderitz & Bauer, Berlin
2152/3020-543210

Editorial Board

Table of Contents

Chemical Fossils in Early Minerals

Hans D. Pflug

Geologisch-Paläontologisches Institut, Justus-Liebig-Universität,
Senckenbergstr. 3, D-6300 Gießen, FRG

Table of Contents

Topics in Current Chemistry, Vol. 139
© Springer-Verlag, Berlin Heidelberg 1987

I Introduction

Precambrian sediments offer a continuous record of reduced (organic) carbon from their first appearance 3800 my ago (Schidlowski et al., 1979)[1]. Most of this sedimentary organic matter is preserved in the form of kerogen which is defined as the acid insoluble, polycondensed end product of diagenesis of organisms and their metabolic products. The ultimate source of this reduced carbon is, therefore, photosynthetic activity by plants, since low temperature reduction of oxidized carbon (CO_2, HCO_3^-) in terrestrial near-surface environments is primarily the result of photosynthetic carbon fixation (Schidlowski, 1984b)[2]. One major aim of organic geochemistry is to isolate and identify chemical fossils from the kerogenous material. The term chemical fossils was originally introduced to describe organic molecules that survive the passage of time with little or no alteration in their basic skeleton. Carbon-carbon bonds are fairly resistant to thermal cracking, and particularily hydrocarbons possess a considerable preservation potential surviving as molecular fossils even in sediments of the Archean and lower Proterozoic dating back 3800—2000 millions years (= 3.8—2.0 Ga) ago. Organic remains from this time span are the main subject of the present article. The reviewed literature has been surveyed up to April, 1985.

The biosynthesized lipids and pigments in particular have characteristic carbon skeletons which are preserved in the form of saturated or aromatic hydrocarbons under favorable geological conditions, and may thus qualify as biological markers. However two severe problems arise in the organic geochemistry of very old rocks.

High temperature metamorphism can alter the kerogen until only carbon-carbon bonds are present in the form of amorphous carbon or graphite. Consequently, the molecular information is completely lost, and isotopic information may be obscured as well (Eglinton, 1983)[3]. Pyrolysis techniques which are commonly used for the analysis of ancient geopolymers have often yielded controversial results. Leventhal et al. (1975)[4] pyrolised 30 Precambrian kerogens and concluded that no biochemical fossils could be recovered from this ancient organic matter. But pyrograms published more recently from these kerogens show typical features of algal-type material with constituents similar to isoprenoid structures (McKirdy & Hahn, 1982[5], Philp & Van De Ment, 1983)[6]. With such findings another problem requires consideration. Geologically younger organic compounds may have contaminated the Precambrian sediments, because most rocks are sufficiently porous and permeable to groundwater.

In order to overcome these problems, interest was focussed on that portion of the organic matter trapped in mineral precipitates which formed synchronously with sedimentation. In these cases, the material is hermetically sealed in the crystalline matter and may survive with relatively little subsequent alteration. Such preservation is common in cherts which are chemical precipitates of silica and now consist of fine grained quartz. These rocks offer the best chance for successful preservation of truely Precambrian molecular fossils. Modern microprobes and spectrophotometer microscopes allow the non-destructive analysis of organic matter enclosed in mineral crystals. Laser bombardment of microscopic sites deliberates molecules or molecular fragments that are subsequently passed through a mass spectrometer.

The major goal of such analyses is to correlate the morphological evidence of microfossils sealed in crystals with chemical data. This is desperately needed because the microfossils preserved in ancient sediments are commonly not more than simple spheres or filaments lacking the morphological details characteristic of more recent fossils. They often resemble bacteria or other unicellullar organisms, but it is often debatable if such simple structures are distinguishable from similar forms of non-biological origin. It is surprizing to note, that, although thousands of such specimens have been studied, described and interpreted as evidence of past life, few descriptions give even rudimentary information on their chemical composition. In most cases it is not even known if organic material is contained in the structures. But in the absence of definite ultrastructural and microchemical evidence it is often difficult, if not impossible to identify them as biological entities. Consequently, the most convincing way of assessing the biogenicity of such fossil finds is by relating morphology and chemistry. What can result from the studies is a set of criteria, a listing of what might be termed "reasonable attributes of ancient microfossils" [7] (Schopf & Walter, 1983).

A most remarkable result in the chemical study of ancient morphologies is the complex chemical composition, comprizing carbonaceous materials and different inorganic materials such as carbonates, iron sulphides and iron oxides. It can be shown that most of the inorganic constituents are "biominerals" produced directly or indirectly by the life activity of the involved organisms. So also these products are chemical fossils and, therefore, merit adequate consideration in our analyses.

A special reason for the expansion of interest in the inorganic chemistry of fossil microbes results from the evidence that many of them precipitate economically valuable minerals, either actively or passively.

The processes leading to the fossilisation of such microbes are important aids in understanding the genesis of sedimentary ore. The association of organic matter with Precambrian ore deposits is relatively common but inadequately documented, and its genetic implications are still poorly understood.

II Permineralization of Biological Matter

The best preservations of organic particles are found in quartz grains of syn-sedimentary origin. It is probable that the material was entrapped in amorphous silica which, upon dehydrating to solid opal, formed an incompressible matrix with minimal deformation. The resistance of opal and of the subsequently crystallized quartz provides a physical environment which preserves the structures and reduces the effects of heat and pressure over long periods, which otherwise lead to degassing and coalification of the organic matter.

The source of the large amounts of silica present in Precambrian cherts is unclear. Since few siliceous organisms (Klemm, 1979) [8] have been reported to occur so early in sediments, it is assumed that most or all of the cherts were of non-biological origin and derived as precipitates from silica-saturated waters. The mechanism involved is not very well understood, but it appears to be

(1) rather rapid,
(2) related to rather sudden environmental alterations such as p_H-alteration and
(3) mostly associated with abundant organic material (Huestis, 1976)[9].

Immediate replacement seems to occur by means of chemical diffusion in carbonate material (Schmitt & Boyd, 1981)[10]. Decay of proteinaceous compounds, decarboxylation and/or fermentation of associated organic matter may have formed the microenvironments in which the p_H was lowered. Adjacent calcite then begins to dissolve. Dissolution and silica precipitation appear to take place on a microscopic scale along solution films. The calcite releases Ca^{2+} and HCO_3^- into the solution film. Such pore waters contain much H_4SiO_4, the dominant silica species in near neutral and acidic solutions, but relatively little Ca^{2+} and HCO_3^-. Consequently, these latter ions move by diffusion into the pore waters. The solution film, on the other hand, contains less H_4SiO_4 than the pore waters, causing diffusion of H_4SiO_4 from pore waters to solution films (Fig. 1).

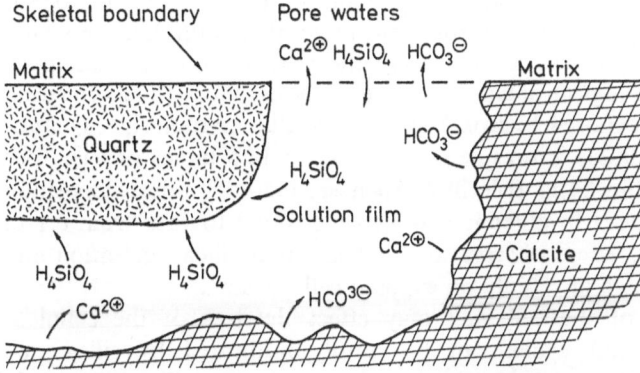

Fig. 1. Replacement of calcite by SiO_2 in a fossil. Diagrammatic relationship between solution films and pore waters. Direction of movement of chemical species indicated by arrows (Schmitt & Boyd, 1981)[10]

In other cases, silica mineralization is a mere impermeation or void-filling process. It has been observed that silica was deposited in all cracks and openings between associated plant cells and in the spaces left by cell fluids. Sigleo (1978)[11] analyzed cross sections of silicified wood with an electron microprobe (SEMQ) and compared the distributional pattern of silica, carbon and calcium with the cell morphology. He found that the calcium patterns provided the best structural information because the distribution of calcium precisely reflects the cell wall features in the carbonaceous sample. His scanning electron micrographs clearly illustrate the void-filling nature of the silica mineralization. The H/C atomic ratio in the carbonaceous material were found to be lower than that of modern lignin, indicating a higher degree of aromaticity in the former. IR spectra indicate a loss of functional groups, in particular methoxy groups. The main absorption bands for silica filled wood are in the regions of 3450 (aromatic-OH),

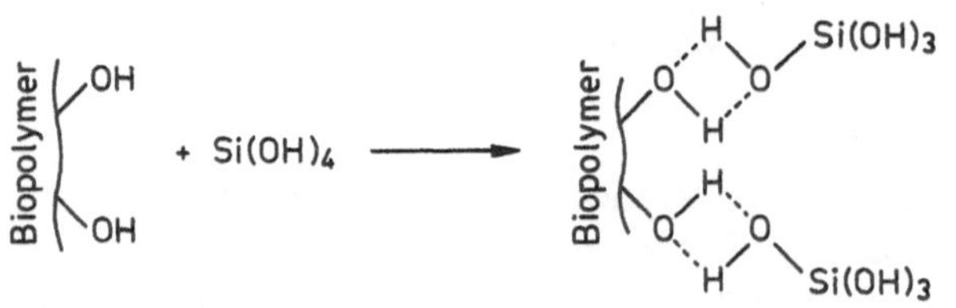

Fig. 2. Suggested interaction of silicic acid with organic matter through hydrogen bonding (Francis et al., 1978) [15)]

1605 (aromatic C=C) and 1080 (Si—O—Si). The silica spectrum can be eliminated after treatment with HF.

Leo & Barghoorn (1976) [12)] suggested that the emplacement of silica during petrification may involve the establishment of actual bonds between the different materials (Fig. 2). They point out that the polysaccharids (including cellulose) contain abundant hydroxyl groups which are capable of forming hydrogen bonds. The sheaths of primitive plants such as bluegreen bacteria which seem to become easily embedded in silica contain analogous mucopolysaccharides. It is surmised that $Si(OH)_4$ monomers begin to interact and polymerize, forming siloxane bonds and eliminating water (Fig. 3). It is further surmised that with later polymer growth, silica begins to deposit as a film along the cellular surfaces and in so doing replicates the morphological details of the cell. The cellulose fibrils covering the cell wall contain three exposed hydroxyl groups per glucose molecule. Quantitative data indicate that cellulose fibres can absorp up to 12.5 moles of silica per gram of air-dried pulp within 24 hr (Merill & Spencer, 1950) [13)]. Silica deposition rates on wooden blocks placed in alkaline springs were found to vary from 0.1 to 4.0 mm/yr (Allen, 1934) [14)]. These observations indicate that silica nucleation and deposition can occur directly and rapidly on exposed cell surfaces.

Preceeding degradation of the cell wall may effect the loss of the cellulose leaving a rather porous wall which can be readily infiltrated by silica rich solutions. It is suggested from experiments that the penetrating silica may settle as film also along the interior cell interfaces and interact with the cell content through hydrogen bonding (Francis et al., 1978) [15]. The in vivo pigments present as closely packets of layers at liquid interfaces allow penetration of the silic acid. Chlorophyll exposed to the acid solution soon liberates its magnesium and yields brownish pheophytin.

It should be emphasized that much of the organic material may become decomposed by postmortem processes, both prior to silification and during silification. Francis et al. (1978) [15] reported observations on microorganisms silicified in the laboratory. In some cases, silification altered the sizes and appearance of the specimens severely. Often the cell content collapses and shrinks away from the cell

Fig. 3. Formation of siloxane bonds through polymerization (Francis et al., 1978) [15)]

wall. Consequently most of the inclusions found in silicified cells are degraded and mineralized products.

This brief discussion should emphasize that the interpretation of such an apparently simple process as silification is dependent on the understanding of rather complex chemical systems. More experimental work needs to be done before the processes are adequately understood.

III Analytical Techniques and General Results

III.A Basic Principles

Individual and non-destructive chemical analysis of microscopic remnants enclosed in mineral grains has been proven possible by certain spectroscopic techniques such as Raman-, IR-, UV/visible- and Laser mass spectroscopy (Pflug, 1982) [16].

Table 1. List of applied analytical Techniques

Technique	Instrument	Spot of measurement (μm diam.)
1. Laser Raman	MOLE (J.S.A. Jobin Yvon)	1–5
2. Infrared absorption	NanoSpec/20 IR (Nanometrics)	33
3. UV/visible absorption	UMSP 1 (Zeiss)	1–5
	UV-microscope (Leitz)	
4. Laser Mass Spectroscopy	LAMMA (Leybold)	1–5
5. Electron microprobe	AMR 1600 T/WDX2A (Leitz/MICROSPEC)	1–5

Compared with the usual whole rock analyses, the above techniques have several advantages. (1) The sample can be analyzed as it is, without complicated and time consuming treatments, and a small amount of material is sufficient for obtaining spectra. (2) Not only the carbonaceous matter is analyzed but also the mineral composition, that results from biologically mediated mineral formation (as in iron- or manganese-precipitating bacteria, or in calcified bacterial colonies or cyanobacterial sheaths). (3) Chemical and microscopic analyses are combined. Thus, chemical and morphological findings can be related. (4) The microfossils remain intact throughout the procedure and the analyses are reproducable in local detail. The measurements are not influenced by impurities which often occur in fissures or in pores of the rock.

When used alone however, the methods also have several disadvantages. The amount of information is limited, because scarcely the physical properties alone will give sufficient information to elucidate the chemical constitution of a fossil particle. So it will be mostly impossible to write the structural formula from spectral data alone. But in combination with other data such as those from whole rock analyses, a choice between alternative interpretations is sometimes possible. Often, qualitative information on specific structural elements can be obtained, though the spectra are mostly too complex for individual compound analysis. Another limitation of the above methods is that various types of functional groups contribute to a given absorption band and that the individual compounds necessary for calibration purposes are not readily available. For all these reasons, information is limited to molecular types of hydrocarbons and to structural groups, and it is perhaps in this field that spectroscopy has its most valuable application.

III.B Preparation

A scheme of the technical procedure that has proven useful for the concerning studies is shown in Fig. 4 (Pflug, 1984a) [17]. Compact and unweathered rock samples are selected for the analyses. The rock sample is sectioned into two halves (top of Fig. 4). From one side of the main cut (1 b) a normal ca. 30 μm thin section is produced. From the opposite surface of the cut (1 a), a thick section is prepared for the purpose of demineralization. The thin section (1 b) is used for polarization microscopy and microprobe analyses.

Under the light microscope, the thin Section (1 b) reveals the original condition of the particles, with respect to their distribution and arrangement within the mineral matrix. Special note must be given to open cracks, since they might contain contaminants (Fig. 5). It is possible by localized comparison of the two rock sections 1a and 1b to detect the morphological changes that take place during the demineralization process. In order to avoid such problems, in situ demi-

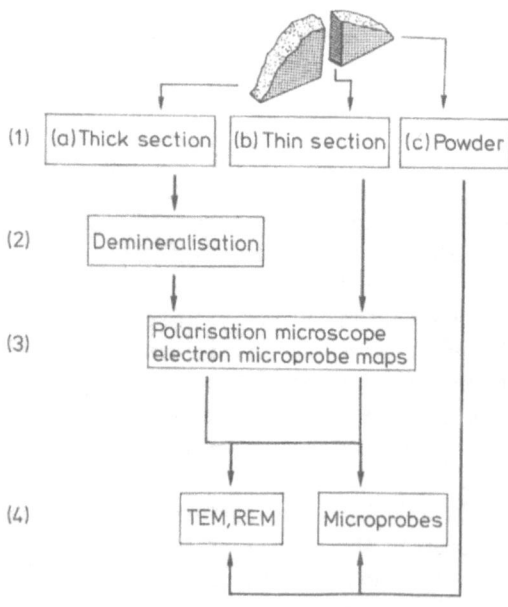

Fig. 4. Preparations and analytical procedure (Pflug, 1984a) [17]

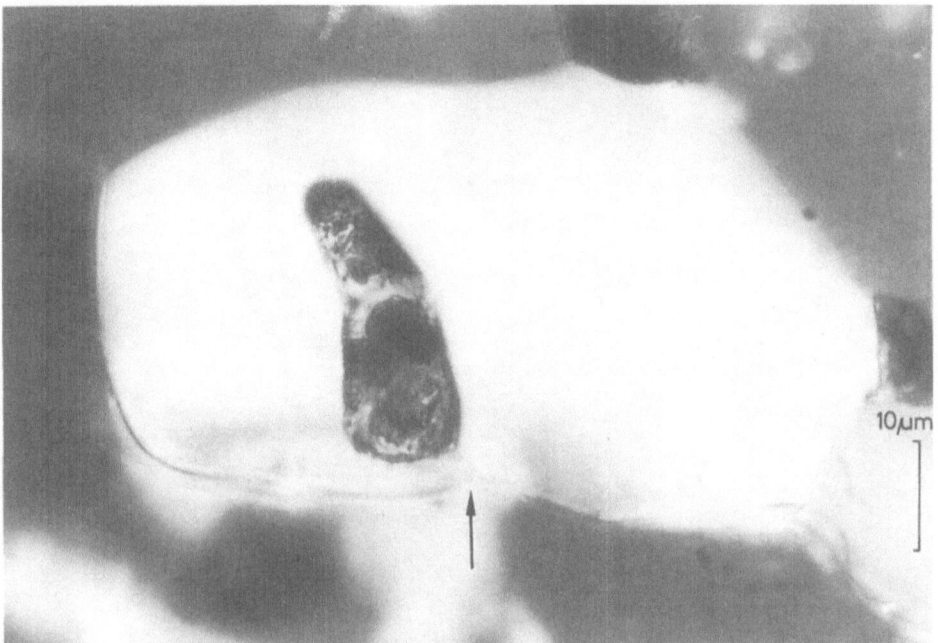

Fig. 5. Carbonaceous inclusion in a quartz grain. Carbonaceous material partly leached out by weathering through microfissure (see arrow below) (Pflug, 1985) [18]

neralization is carried out on a membrane filter. A section of the rock ca. 1 mm thick is placed on a membrane filter of 0.01 μm pore diameter. It is then exposed to vapors of HF and HCl. As a result of the treatment, the minerals present (or most them) are dissolved and removed through the pores of the filter. The organic particles are not affected by the procedure and remain on the filter with no or very little alteration of their original structure and position (Fig. 6). The demineralized section can be immediately examined under the light microscope together with the underlying filter as a support. Preparation for the TEM is not complicated.

Stricktly speaking, kerogens are always inhomogeneous and always contain "impurities" such as graphite particles, "carbon-black" carbonaceous particles at various degrees of metamorphic alteration and mineral impurities. An electron microscope is a particularily suitable tool for studying these phases present in kerogens, since particles less than 1 μm can be examined (Oberlin et al., 1980) [19].

III.C Infrared and Laser Raman Microspectroscopy

Computerized IR microspectrophotometers such as the Nanospec (Table 1) can scan areas as small as 20 × 20 μm over the infrared wavelength range of 2.5 to 14.5 μm (4000–690 cm^{-1}). Samples to be measured are placed on the microscope stage. Selection of small areas is possible by direct viewing at 150 × magnification.

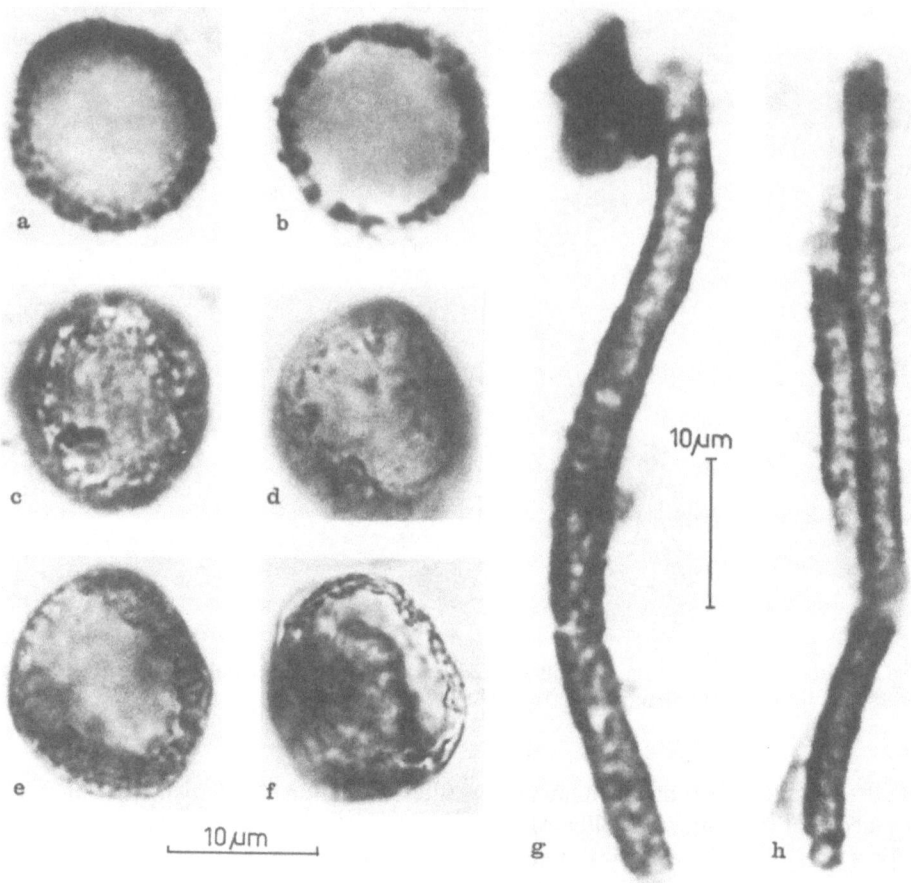

Fig. 6a–h. Spherical and filamental organic microstructures, **a–c)** *Huronispora* (Gunflint), **d–f)** *Isuasphaera* (Isua), **g, h)** *Gunflintia* (Gunflint). Figs. **a–c), g, h)** = demineralized condition, Figs. **d–f)** = in thin Section. Bar for Figs. **a–f)** below, for Figs. **g, h)** between the Figs. (Pflug, 1985) [18]

Several reviews are available on the infrared spectroscopy of fossil organic materials (Speight, 1971 [20], Robin et al., 1977 [21], Rouxhet et al., 1980 [22]), but no detailed objective review exists on the application of such spectroscopic techniques on individual particles of microscopic size.

Comparable infrared (IR) spectra of complex organic solids such as coals (Fig. 7) [23], cherts, kerogens, humic substances and some natural polymers have been presented in various publications. They show a limited number of rather broad bands which are due to well defined chemical groups and can often be interpreted by comparison to less complicated spectra. The signals commonly observed in fossil organic matter are as follows (Robin et al., 1977 [21], Tissot & Welte, 1978 [24], Rouxhet et al., 1980 [22], Friedel & Carlson, 1972 [25]):

(1) A broad absorption around 3430 cm^{-1}, attributed to OH groups.

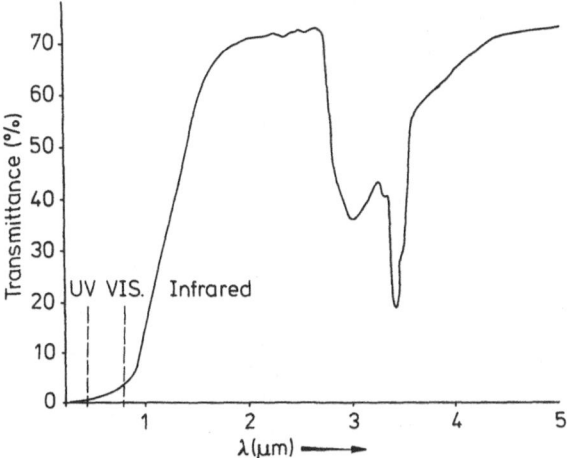

Fig. 7. Electronic absorption shown by 20 μm thin section of Pittsburg seam anthraxylon (Friedel, 1960) [23)]

(2) Absorption bands in the typical C—H stretching region at 3000–2870 cm^{-1} usually showing two maxima at 2925 and 2855 cm^{-1} related to CH$_2$ and CH$_3$ aliphatic groups. These bands are well defined in the spectra obtained from microbial biomass of soils and there can be assigned to cell wall and capsular polysaccharides (Filip, 1978) [27)]. But they are also common in spectra of other biological materials.

(3) A wide band with a maximum around 1720 cm^{-1} which is assigned to various C=O groups. Carboxyl groups of lipids as well as ester carbonyl groups absorb very strongly at 1730 cm^{-1}. Additionally, COOH-groups of undissociated amino acids may participitate in the absorption of that region.

(4) Broad absorption bands with a maximum around 1630 cm^{-1}, usually related to aromatic C=C, although there is definitely some contribution from other structures.

(5) A wide band between 1400 and 1040 cm^{-1} which implies C—O stretching and OH bending.

(6) A succession of weak bands from 900 to 700 cm^{-1}, related to various aromatic CH (bending out of plane) and dependent on the number of adajcent protons (Fig. 8).

Information is drastically reduced when the particles are analyzed in situ in the crystal. Except for C—H stretching bands in the 2950–2855 cm^{-1} region, no distinct organic absorption bands are present at wave lengths shorter than 8 μm wavelength. The absorptions of Si—O, O—Si—O and Si—O—Si atomic groups of the mineral matter are sharply produced, indicating quartz and other silicates. In the region of 1000–690 cm^{-1}, the absorption bands of metal oxides (Al, Fe, Mg)—O—O and the —CO$_3^{2-}$ group of carbonates appear. The various carbonates can be readily identified in the Raman- and IR-spectra. In the latter, a carbonate absorption band often occurs at 1420 cm^{-1}, two weaker bands lie at 855/865 and 710/730 cm^{-1}. Sharp bands at 425 and 350 cm^{-1} are due to pyrite. Iron hydroxides appear in the IR-spectra with broad absorption around 3140 cm^{-1} together with two other bands at about 900 cm^{-1} and 800 cm^{-1}, respectively. General data on the

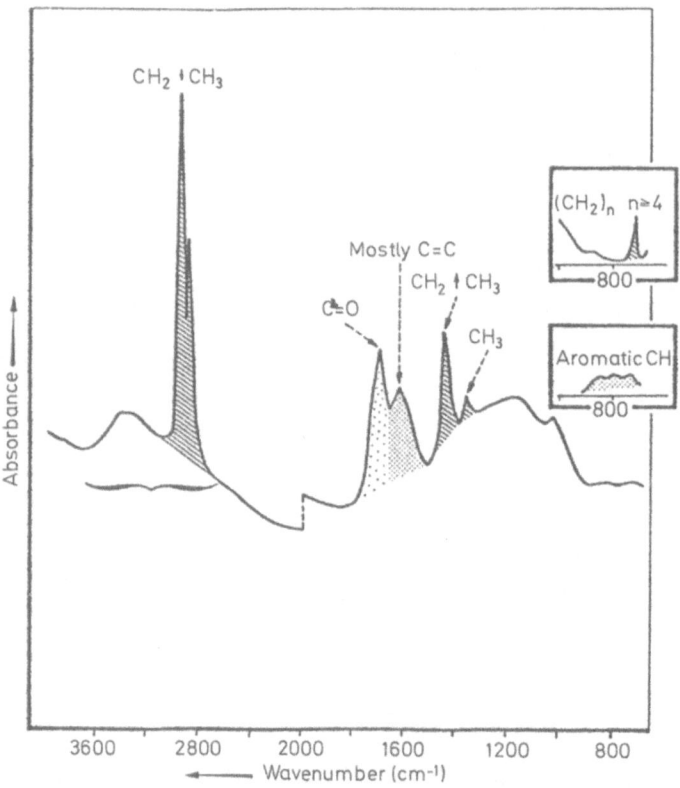

Fig. 8. Typical IR spectrum of microbial (type II) kerogen. The two insets show specific absorption bands of two other kerogens: $-720\ cm^{-1}$ related to long aliphatic chains mostly occurring in type-I kerogen -930 to $700\ cm^{-1}$ related to aromatic CH (bending out of plane) mostly occuring in very mature kerogen (Adapted from Robin[26], 1975, in Tissot & Welte, 1978)[24]

infrared absorption and Raman spectroscopy of inorganic constituents of rocks can be found in Hunt et al. (1950)[28], Chester & Elderfield (1968)[29], Liese (1975)[30], Griffith (1975)[31], Van der Marel & Beutelspacher (1976)[32].

A limited number of parameters calculated from the absorption coefficients has been proven useful. Although they are empirical to some extent, they have a rather clear chemical meaning. In fact, the spectra differ from each other by the intensity of the various bands. For instance, the spectra of kerogens and coals vary in regular fashion according to rank and reflect the main chemical modifications occuring as rock metamorphism proceeds (Rouxhet et al., 1980)[22] (Fig. 15, p. 23).

The band intensities arising from the aliphatic C—H bonds are of particular interest, since they depend on the atomic weights of the atoms to which the other three valences of the carbon are linked. The peaks around 2930 and $2860\ cm^{-1}$ are due mainly to the asymmetric and symmetric stretching of alkyl CH_2 groups, expected at 2926 and $2853\ \pm\ 10\ cm^{-1}$. The CH_3 groups are expected to give asymmetric and symmetric stretching bands at 2962 and $2872\ cm^{-1}$.

They are responsible for broadening of the band at high wavenumbers, which sometimes appears as a shoulder at 2960 cm^{-1} in coals and kerogens with little hydrogen. The absorption due to the symmetric stretching of CH_3 groups is responsible for a decrease in the valley separating the two peaks. The relative abundance of CH_3 vs. CH_2 groups is reflected by the half band width of the peak at 2930 cm^{-1} and by the depth of the valley separating the peaks at 2930 and 2860 cm^{-1}. Such parameters are not very accurate and can only be used to detect trends among series of samples.

The Laser Raman microprobe constitutes a physical method of microanalysis based on the vibration spectra characteristic of polyatomic structures. A focused laser beam excites the sample. The light diffused by the Raman effect is used for identification and localisation of the molecular constituents present in the sample. An optical microscope allows a survey of the interesting structures and the placing of the laser beam. The spectra obtained from fossil organic particles generally match well the corresponding IR-spectra, but the features in particular yield additional information, which will be discussed below with the given examples (Fig. 23, p. 36).

III.D UV-visible Microspectroscopy

UV-visible-microspectrophotometers consist basically of a microscope containing quartz optics and illuminated by changing wavelengths of light. The ultraviolet and visible absorption spectra of particles as small as 1 μm in diameter can be determined by this technique. Thin Sections as well as demineralized sections of the rock are used for the studies. The method is a highly sensitive one capable of detecting many compounds in the ppb range. But the ultraviolet spectra of organic molecules are often difficult to interprete.

A frequent approach for investigating the properties is to study comparable compounds. Sporopollenin, a material isolated from pollen or spores is considered to be an interesting model compound of the parent organic matter that leads to formation of kerogens with a large number of aliphatic groups. Many sporopollenins show an absorption maximum at 280 nm which among others is characteristic of both five- and six-membered rings with conjugated double bonds (Fig. 9). Other classes of compounds with absorption maxima in the same region as sporopollenin include dienones (290, 292, 315 nm), especially those with hydroxyl group substitutions, and polyene acids (294 nm). These types of bonds may be present in carotenoid ester units (Southworth, 1969)[33]. Through fossilization, the original compounds may interact with each other to change the fine detail of the absorption spectra. A large amount of nitrogen and sulphur in sediments is present in the form of decomposition residues which contain the chromophoric groups R—CO—NH—R', R—S—CO—R', and others, causing absorption in the ultraviolet. N-Heterocyclics and aromatics consisting of one or more rings may also be present in smaller or larger amounts.

All known spectra of fossil organic materials have certain features in common (Fig. 10). Absorption increases gradually from the infrared through visible and ultraviolet to about 220—240 nm. The only structures regularly occuring are several absorption bands or a broad shoulder between 260 and 280 nm superposed on a broad background absorption (McCartney & Ergun, 1967 [34], L. A. Nagy, 1978) [35]. The structureless absorption may be partly caused by light scattering and partly due to electronic absorption of condensed aromatic rings. Similar spectra are known to exist in coals and coal derivates as well as in chars prepared from oxygen containing compounds such as carbohydrates (Friedel & Retcofsky, 1970 [36], Retcofsky & Friedel, 1970) [37].

Fossil-organic particles in situ are often encrusted or impregnated by iron compounds. Fe^{3+}-oxides and -hydroxides appear with two strong charge transfer bands in the UV-visible absorption spectra, one at 395 nm, the other at 465 nm

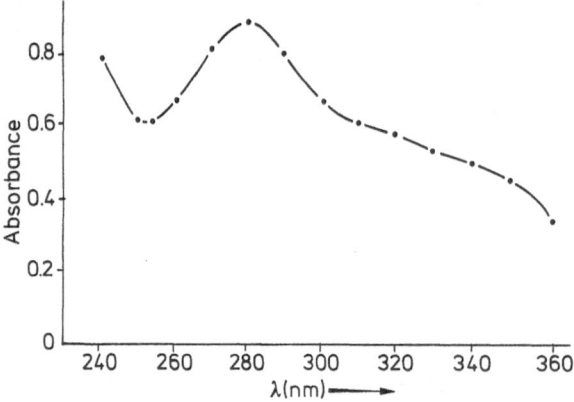

Fig. 9. UV-absorption spectrum of recent spore *Lycopodium taxifolium*, distal wall (Southworth, 1969) [33]

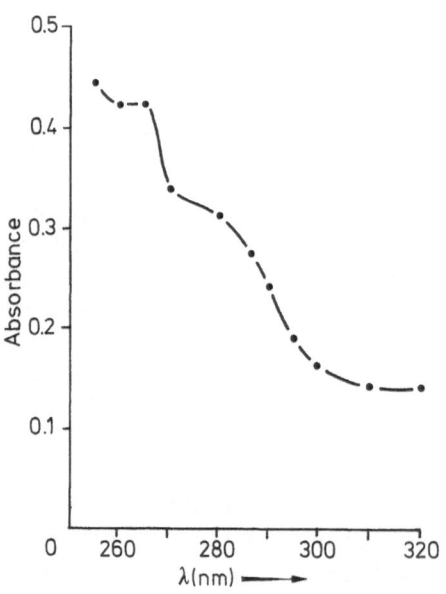

Fig. 10. UV-absorption spectrum of a filamentous organic microstructure from Transvaal stromatolite (ca. 2200 MY). Beam diameter of the instrument was below 1 μm. Absorption shoulders at 280 nm and 265 nm were assigned to nitrogenous organic compounds containing double bonds (L. Nagy, 1978) [35]

plus two ligand field bands at 650 nm and 900 nm respectively (Bell et al., 1975) [38]. The strong signals can mask the organic functions of the spectrum completely. In this case, the organic particles need to become demineralized before measurement.

Summed up it can be stated that the UV-visible microspectra are not definitive enough to satisfactorily establish the chemical nature of the particles, but the value of the technique increases when combined with other analytical methods. UV visible absorption has shown to be especially informative in the study of metamorphic alteration of organic matter (see chapter V).

III.E Laser Microprobe Mass Analysis

Laser microprobe mass analyzers permit mass spectrometric analysis of very small volumes (0.01–1 μm^3) of thin Sections. The method is based on laser induced ion production from a microvolume and analysis of the evaporated ions in a time-of-flight mass-spectrometer. The technique allows detection of all elements and isotopes with a sensitivity approaching the ppm range and an extremely low limit of detection: 10^{-15} to 10^{-20} g. Transmission type instruments such as the LAMMA 500 are designed for the analysis of particles of $\gtrsim 3$ μm in diam. The lateral resolution is about 0.5–1 μm. Because the area to be analyzed is selected by an optical microscope, distribution of chemical constituents can be precisely correlated with morphologic structures (Hillenkamp et al., 1982 [39], Simons, 1984 [40], Kaufmann, 1984) [41].

Thin Sections or pieces of the rock about 1 mm in size or less can be also analyzed with this instrument at glancing incident irradiation, if the site to be analyzed is close to the surface of the section or the edge of a fraction. However in this mode of operation, mass resolution is severely reduced. Nevertheless ions can be identified because of their low mass numbers and the very low background in the spectra, by known cluster patterns or by comparison with reference spectra obtained with reduced laser irradiation. The spectra have shown to be highly reproducible. For more precise bulk sample analysis, other instruments such as the LAMMA 1000 are more suited. Their lateral resolution is 1–3 μm in diameter and the depth of analysis is typically 0.1 μm.

Fossil organic particles in situ usually release positive ions at masses 23, 24, 39, 41, 54, 56 thus indicating Na, Mg, K and Fe. But other signals also occur (Fig. 19). The negative ion mass spectra are dominated by carbon clusters with zero, one or two hydrogen atoms attached. They resemble spectra obtained from polymer foils (Gardella et al., 1980) [42] rather than those from coals and carbon films (Fürstenau et al., 1979) [43]. Consequently, the material contains more long hydrocarbon chains rather than aromatic constituents. Peaks at $m/e = 79$ (benzylium) and 90 (tropylium) indicate aromatic constituents. Unspecific ions like CN, CNO, and Cl are commonly present (Fig. 11).

In the fossil materials studied, the $C_n H_m$ clusters exhibit high intensities around $n = 3, 4, 5$ or 6 and then decline in intensity (Fig. 11). The highest negative mass ion detected was at 156 m/e. Small peaks produced at higher laser energies, could result from rearrangements during the ionization process, when a laser

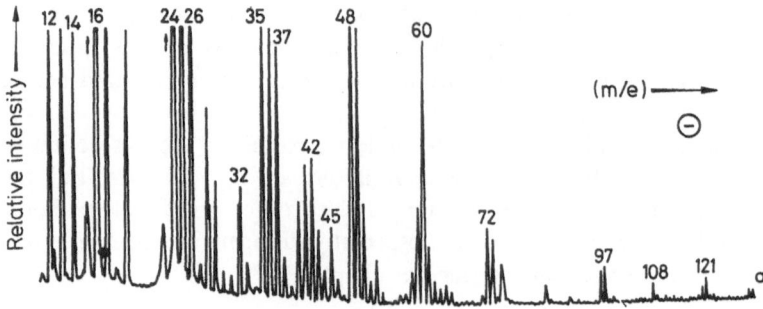

Fig. 11. Laser mass spectrum (negative ions) from cell wall of *Huronispora* included in thin section of the Gunflint chert (compare Figs. 6a–c) Field of measurement ca. 1 μm diam. (Pflug, 1982) [16]

Table 2. Laser mass spectrum (negative ions), Attribution of signals

C_n^- and $C_nH_m^-$ Clusters:	12, 24, 36, 48 ...: C_1^-, C_2^-, C_3^-, C_4^-, ...
	13, 25, 37, 49 ...: CH^-, C_2H^-, C_3H^-, C_4H^-, ...
	14, 26, 38, 50 ...: CH_2^-, $C_2H_2^-$, $C_3H_2^-$, $C_4H_2^-$, ...
Others:	16: O, 17: OH, 19: F, 26: CN^-, 28: Si, 32: S^-, 35, 37: Cl^-,
	40–42, 45: Organic fragments, 60: SiO_2^-, 76: SiO_3^-

induced plasma exists for a short time. But these phenomena are minor to the production of molecular fragments directly related to the structure (Gardella et al., 1980) [42].

IV Host Sediments

IV.A Banded Iron Formations (BIF)

Stromatolites and Banded Iron Formations (BIF) are the main host sediments of chert-bound chemical fossils. The BIF characteristically consist of alternating laminations of chert and iron minerals such as magnetite (Fe_3O_4) or hematite (Fe_2O_3), iron carbonates and silicates. BIF deposition is typical of the Precambrian and does not occur in subsequent geologic ages. The facies makes its appearance at the beginning of the rock record at about 3800 MY and thus belongs to the oldest sediments known on Earth.

Several hypotheses of the possible relationship between the deposition of the cherty iron formation and the activity of various primitive organisms have been defined and repeatedly proofs have been offered to support the occurrence of definite biota. In fact, some indication is available for the assumption, that life was abundant at the time and place of deposition of the BIF and that iron formations themselves were deposited as a result of biological processes. It is presumed in this context, that iron-bacteria reacted with the oxygen acceptor Fe^{2+} in solution and then deposited the trivalent and/or trivalentdivalent iron as precipitated compounds along with other residues resulting from biomass. All these components were then eventually interspersed with chemically formed silica gel (Melnik, (1982) [44]), (Figs. 35, 36, p. 45–46).

Fe and Mn precipitating bacteria are known to occur in many diverse environments, both modern as well as ancient (Muir, 1978) [45]. More recent data presented by Cowen (1983) [46] demonstrate that Fe and Mn precipitating bacteria are also present in suspended pelagic particulates suggesting active biomineralization at these sites. Their study was made possible by specialized methods for the collection of fragile macro-particulates suspended in the ocean and by the Scanning Transmission Electron microscopy (STEM)-Energy dispersive X-ray spectroscopy (EDS) analysis technique. The use of thin sectioned material and the STEM-EDS allowed the simultaneous recording of the internal ultrastructure and the identification and microanalysis of submicrometer particles including polymer enshrouded bacteria.

IV.B Stromatolites

Carbonate rocks containing laminate structures in the form of domes or columns are called stromatolites. As can be seen from modern stromatolites, such rocks are mainly produced by $CaCO_3$-precipitating and sediment-binding bacteria. Bacterial mats of this type make their appearance at the beginning of the unmetamorphosed rock record at about 3.5 Ga and still occur today. Biological production, lithification and cementation in these environments have been studied extensively (Golubic, 1984 [47], Awramik, 1984 [48], Monty, 1984 [49]). Lithification stages of bacterial cells can be demonstrated in field samples by combining SEM with Electron dispersive X-ray spectroscopy using spot analyses, linescan or distribution maps (Krumbein, 1978) [50].

The observations have shown that consideration must be given to several different processes of carbonate deposition and/or silica or iron oxide deposition in contact which such bacterial mats. Obviously some important lithification processes take place within the decay zone below the active photosynthetic zone. In most of the cases where lithification was observed there, it was carbonate lithification of a type not related to the photosynthetic depletion of CO_2. Different filamentous and coccoid cyanobacteria can become more or less lithified depending on slime production, mobilization, outer morphology and microenvironments.

Pore water data have shown that the oxidation of the cyanobacterial mat by anaerobic bacteria is a major process leading to the precipitation of $CaCO_3$ (Lyons et al., 1984) [51]. In this process, SO_4^{2-} is removed from the water by sulfate reduction, and Ca^{2+} and biologically produced HCO_3^- are deposited as authigenic $CaCO_3$. Correspondingly, the microbial mats can be considered as a two-layer structure. The upper layer is dominated by filamental or unicellular photosynthesizers and the lower-one by sulfate reducers. Other processes may also be involved in the formation of carbonates (see chapter VII.B).

According to observations on recent occurrences, a relatively high percentage of approximately 2 to 4 % organic carbon of the initial concentration of 16 to 18 % organic carbon is preserved within the final rock generated from the lithified mat. A central topic in this context is the tracing of molecules which survive the decomposition cycles and become chemical fossils. New indication may become apparent when macromolecular complexes such as the microscopically visible capsular slimes and sheath materials are structurally characterized. The glycocalyx of various microorganisms seems to have an excellent preservation potential in view of the presence of sheath like structures in many stromatolites. But information about the chemical composition of the glycocalyx or outer sheath of cyanobacteria is still very limited. Ultrastructural studies point to a rather complex microfibrillar network containing larger fibrils. Cytochemical tests suggest the presence of acidic mucopolysaccharides. Sheath material isolated from *Anabaena cylindrica* consisted of 66% carbohydrate (glucose, mannose and xylose as major components) and 5% amino compounds. A sheath fraction isolated from *Chlorogloeopsis* contained 38% (w/w) carbohydrates and 22% (w/w) protein (Schrader et al. in Boon, 1984) [52].

Capsular slimes and sheaths are a very conspicuous feature of microbial mats. Boon (1984) [52] has succeeded in isolating a few milligrams of sheath tubes from

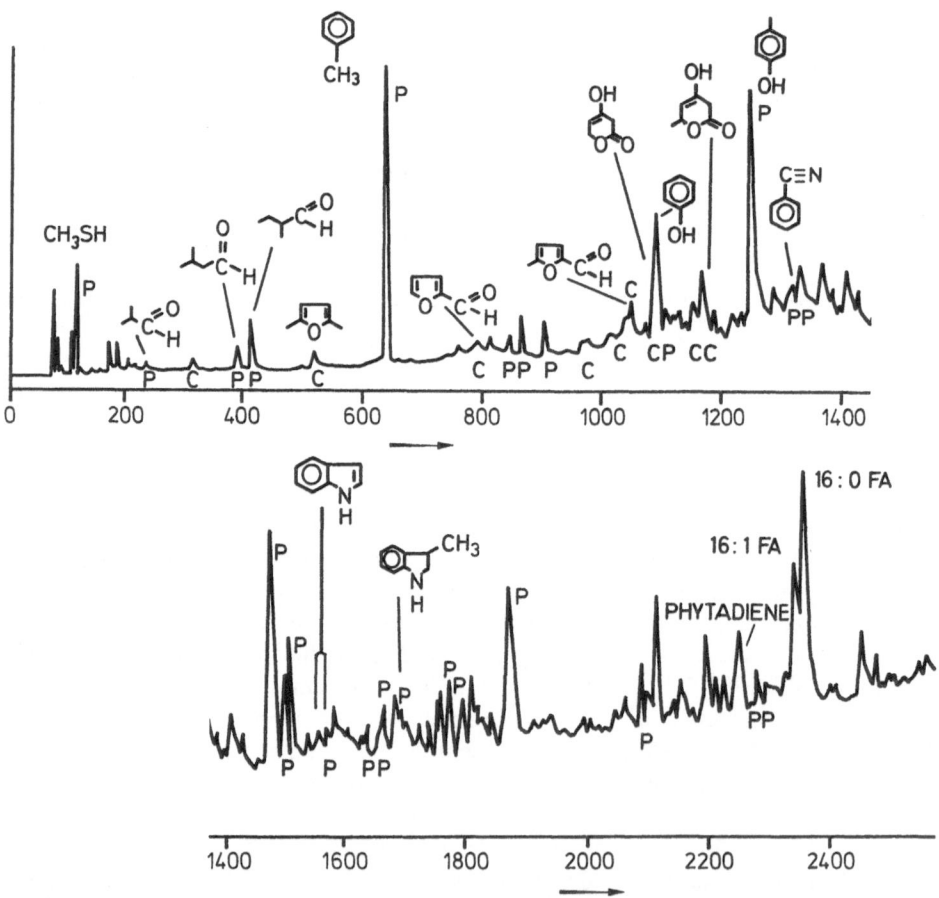

Fig. 12. Total ion current trace of the Curie-point pyrolysis gaschromatography mass spectrometry run of fossil sheath tubes isolated from the Solar Lake cyanobacterial mat sediments (Boon, 1984) [52]

a fossil microbial mat and analyzing them by pyrolysis methods (Fig. 12). Peaks marked with C and P are known to be products originating from carbohydrates and proteins. The carbohydrate fraction is represented by several unspecific furan compounds and two 3-hydroxy-penteno-lactones, which point to the presence of pentose (xylose?) and deoxyhexose. It is suggested, that in fossil condition, only a biodegraded sheath fraction remains, which is depleted in carbohydrates and mainly consists of a residual protein-carbohydrate complex. The findings are of special interest because this type of microbial mats can be conceived of as living precursor to microfossiliferous chert stromatolites and perhaps other shallow water laminated cherts of the Precambrian.

V Organic Metamorphism

It is of utmost importance to understand the metamorphic history of rocks in which the organic matter is formed because temperature, pressure, etc. may alter or even obliterate organic matter and microfossils during metamorphism. The term organic metamorphism is applied to organic transformations occuring under conditions corresponding to incipient greenschist facies metamorphism (> 250 °C). But the term is also frequently applied in the American literature to all the modifications during burial regardless of temperature. Organic matter progressively buried in sediments is degraded. As the temperature increases tars are released, then gases. Simultaneously the carbon content of the kerogen left behind increases until it is almost pure carbon which can eventually be transformed into graphite (Durand, 1980) [53]. The conditions that lead to the dehydrogenation of the kerogen must also affect, to a greater or lesser extent, any coexisting geolipids (i.e., chemical fossils), causing them either to migrate or to become similarly dehydrogenated.

The levels of alteration encountered in Precambrian carbonaceous rock matter are the subject of controversial opinions. It has been suggested by some workers that the age of these occurrences, combined with their degree of metamorphic alteration, has probably totally eliminated the possibility of finding any readily identificable biological markers (Hayes et al., 1983) [54], it was also proposed that Precambrian sediments with low CH_4/C_2H_4 or CH_4/C_2H_6 ratios and H/C ratios > 1 have still the potential to give useful biogeochemical information. Sidorenko (1984) [55] is one of the prominent scientists holding a positive position in the discussion. He concluded from his studies in metamorphic terranes of the USSR, that even up to the amphibolite metamorphic rank, compounds were preserved that had been inherited from the original, probably, sapropelic, organic matter. "A broad range of bituminous components in the originally sedimentary Precambrian rocks, their IR-proved differences of bituminous composition and the composition of gaseous hydrocarbons signify that the original matter must have been different in different regions."

Two points must be taken into account in the present context. Too often kerogen is considered to be a single entity while in fact it is an assemblage of many entities representing different preservational conditions. It must also be taken into account that organic metamorphism in mineral grains cannot be fully compared with that in coal and shales. The latter have porosity sufficient to allow degassing, chemical interaction and homogenization of the organic constituents. These processes are more or less blocked in particles that are hermetically sealed within a quartz grain. So it is by no means clear that metamorphism beyond the greenschist rank necessarily leads to the dissappearance of all chemofossils and to complete graphitization of the organic matter.

What techniques do we have for the study of organometamorphism on a microscopic scale? First there is an approach which essentially uses simple optical analytic methods. It has been proven possible to relate the color of dispersed organic matter under the microscope to the past thermal history of an area or region (Raynaud & Robert, 1976, [56] Peters et al., 1977, [57] Staplin et al., 1982) [58]. Organic particles seen in thin Section can display colors in transmitted light ranging from light yellow to black depending on the extent to which it has been thermally modified. Correlations between the color of amorphous organic matter and many other indices of organic maturity such as the carbon ratios have been reported (Heroux et al., 1979) [59]. The technique has the advantage that the kerogen mixture problem is avoided by measuring light absorption for a single particle. Interpretation of the total kerogen color is often difficult because of the varying maturation rates of the different components present in the sample. Consequently, color interpretations are more accurate when limited to individual particles and when calibrated against other maturation indicators.

Two conclusions can be drawn regarding the use of color measurements for the characterization of Precambrian kerogens: (1) Within an individual quartz grain, the contained carbonaceous microfossils and their associated finely dispersed kerogen are present in comparably preserved condition, exhibiting the same range of yellow to black color; (2) the appearance of some color generally indicates a H/C ratio greater than 0.25.

More elaborated techniques use photometer microscopes with optics capable of measuring the light transmission relative to a blank. The measurements can be combined with thermo-analyses on the microscope heating stage. Hereby, organic particles present in unmounted thin sections are subjected stepwise to increasing temperatures (Fig. 13). Heating causes an alteration of the optical properties as is characteristic of an artifical coalification. Two main processes are indicated in the thermo-diagrams (Figs. 13C, D): Chemical dehydration of the polymer by loss of hydroxyl-groups at low temperatures (1) and a gradual increase in aromaticity at high temperatures (III). Both processes are separated by a phase of plastification (II).

Additional information is obtained from the UV-visible absorption spectra. Coals of low and medium ranks have their main absorption shoulder in the ultraviolet region. The feature gradually shifts to longer wave lengths with increasing rank. It reaches the visible region when the coal has turned anthracitic (McCartney & Ergun, 1967) [34]. Additionally, the materials show a gradual increase in electronic absorption as wave length decreases. Usually the absorptions are higher with increasing rank. The absorption curve for graphite parallels and is reasonably close to those of highest rank. Graphite shows a well defined absorption around 230 to 250 nm. The absorption curves of anthracite and semianthracite are parallel to the graphite curve with the exception of the graphite peak. Measurements on kero-

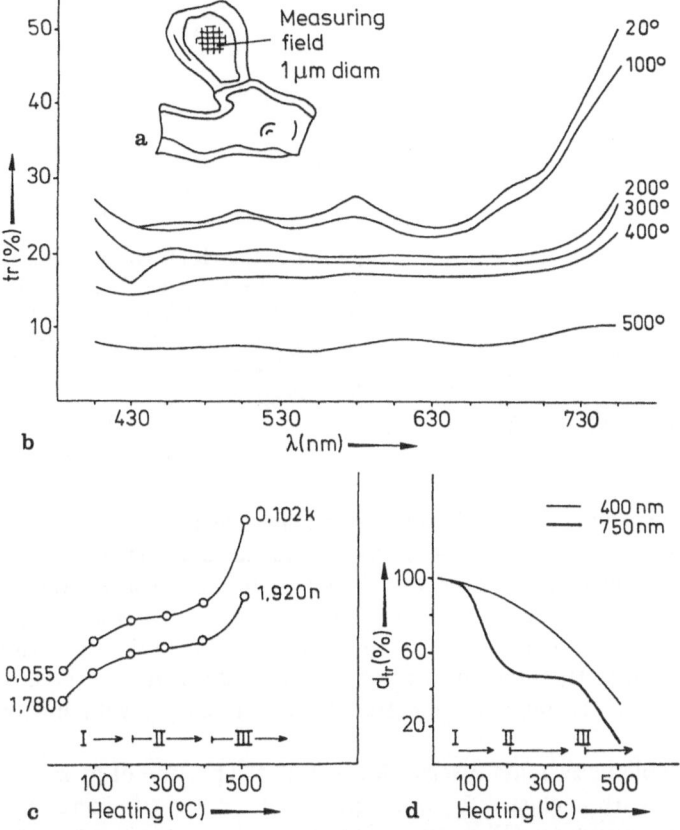

Fig. 13a–d. Thermo-characteristic of organic particle included in Swartkoppie chert. **a)** Location of the measuring spot, cell wall is focussed. **b)** Spectral characteristic of transparency (tr) after gradual heating to 100° ... 500 °C; duration of each heating period: 24 h. The measurings were taken in the intervals between the heating periods, after the specimen has been cooled to room temperature. **c, d)** Alteration of refractive index (n) and absorptive index (k) at 450 nm, and that of relative transparency (d_{t_r}) at 750 nm and 400 nm during gradual heating. I = Phase of degassing, II = Phase of plastification, III = Phase of aromatization (Pflug, 1979) [60]

genous particles produce results that are consistent with and complementary to the above interpretation (Fig. 14).

Infrared and Raman spectrophotometry have become established as worthwile tools in the study of metamorphosed chemofossils. The spectra show which functional groups predominate in immature kerogen and what changes occur during maturation (Fig. 15).

The metamorphic alteration of hydrocarbon structural units has been studied by Robin et al. (1977) [21], who documented the expected disappearance of aliphatic C—H bonds with increasing levels of maturity, and also demonstrated the disappearance of aromatic C—H bonds as structures approach extreme levels of carbonization. In extreme cases when the carbon content exceeds 91% by weight and the H/C atomic ratio is only 0.4, aliphatic and C=O bonds have vanished, and aromatic C=C are the main components remaining in IR spectra. The

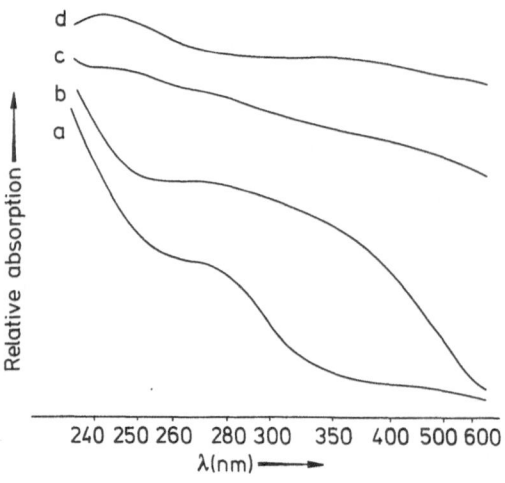

Fig. 14a–d. UV-visible absorption spectra of fossil organic particles after demineralization **a)** Alginite from Permian Boghead coal, **b)** *Huronispora* (Gunflint), **c)** Isuasphaera (Isua), **d)** Graphite (Pflug, 1985) [18]

available data point to the progressive development of a higher degree of order due to increasing polycondensation and more regular spacing of the aromatic layers.

Brown (1955) [61] has studied the IR spectra of coals with special attention on the optical densities of the two peaks at 3030 and 2920 cm^{-1}. He found that the ratio of aromatic hydrogen to total hydrogen increased with rank. Fujii and his colleagues (1970) [62] noted that the absorption band at 2920 cm^{-1} generally increased with rank to 86% carbon but thereafter decreased sharply with higher carbon contents.

Comparison of the available Laser mass spectra shows, that the portion of hydrogen contained in the recorded carbon clusters becomes gradually smaller with increasing rank of the carbonaceous matter (see Figs. 39, p. 49). These are preliminary results which need to be studied in more detail.

There remains one question which is crucial to the subject matter and still awaits critical examination: What happens to an organic particle that becomes sealed in silica and is subsequently dehydrated, fossilized (perhaps chemically bound to the quartz lattice as a result of fossilization), and finally subjected to metamorphism? The temperature-time conditions necessary to convert such a material completely to graphite are not known, not even approximately. But according to the evidence presently available from Precambrian rocks the conditions vary over a broad scale, ranging from medium ranks to graphite. This implies the possiblity of observing fragments which may, at least, be partially derived from biological markers.

VI Contamination

Contamination is a serious problem in the search for chemofossils, but it is not an exclusive problem of the Precambrian. Probably no rock sample on Earth is free of contaminants. Weathered and contaminated rocks may, of course, yield biochemical fossils of undeterminable ages, with the possible presence of modern biochemicals.

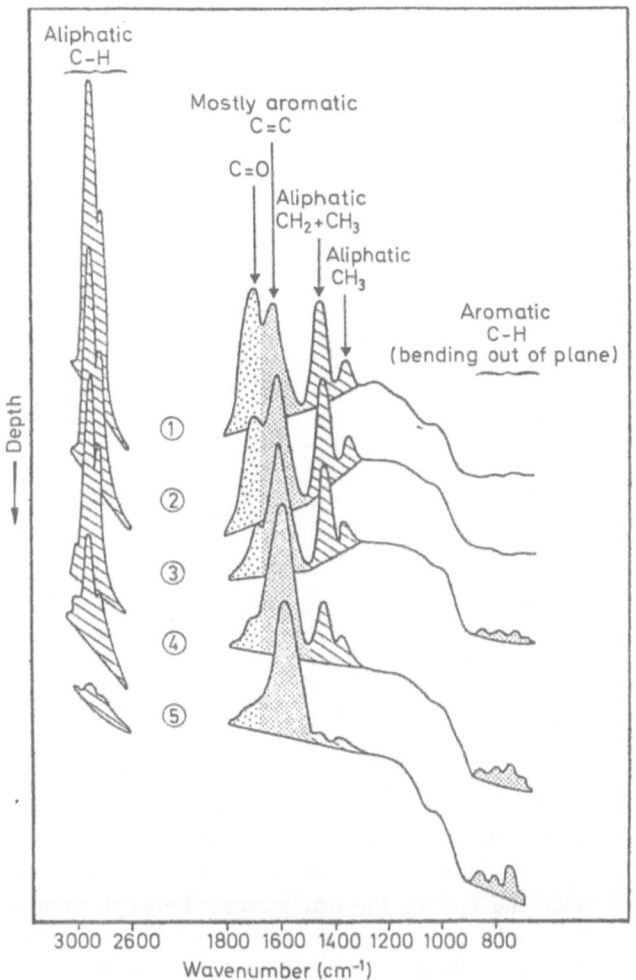

Fig. 15. Infrared spectra showing evolution of microbial kerogen during burial (Tissot & Welte, 1978) [24]

A special problem in this context are the chasmolithic microbes that live within rocks, in cracks and microcavities, especially in carbonate facies (Cloud & Morrison, 1979) [63]. Algae, fungi and oligotrophous nonsporeforming bacteria belonging to the genera *Pseudomonas*, *Corynebacterium* and *Arthrobacter* predominate in weathered zones. Some of them are capable of fixing atmospheric nitrogen, The cenoses include the most primitive, blue-green bacteria (Chroococcales) which are not encountered in most soils. Accumulation of organic contaminants, leaching of indigenous organic material and elements (S, Ca, K, Fe) and formation of clay minerals are typical processes in weathering crusts (Karavaiko, 1978) [64].

Cherts and quartzites are generally believed to be a poor environment for the growth of microorganisms. However, some authors found modern contamination in weathered samples taken from such facies. Nagy et al. (1981) [65] examined

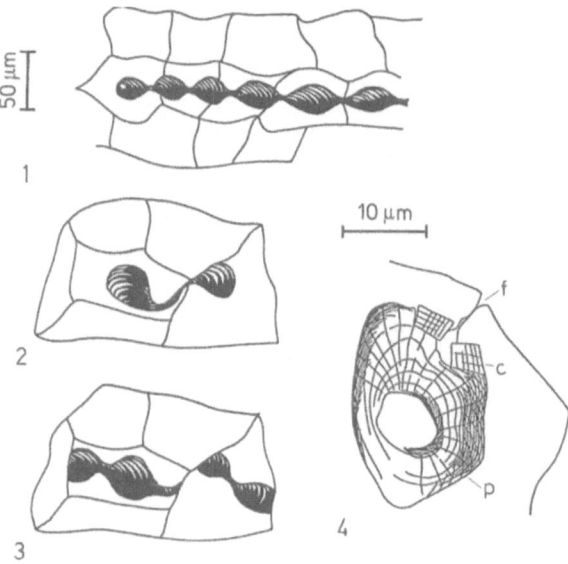

Fig. 16. Post depositional alteration of Isuaspheres included in the Isua quartzite. 1. Before recrystallization of the chert matrix; 2, 3. After recrystallization, 4. Higher magnification of one single Isuasphere that remained as a cavity in the quartz grain after leaching of the carbonaceous matter through the fissure "f", p = imprint of the sheathal structure in the wall of the cavity, c = carbonate. Bar for Figs. 1–3 see left of Fig. 1, for Fig. 4 see above the Fig. (Pflug, 1985) [18]

amino acids and hydrocarbons from the Archean Isua quartzites and concluded, from the extent of amino acid racemization, that there has been an apparent continuous diffusion of biologically derived organic material into the weathering crust from encrusting lichens since the end of the last ice-age. Through cracks, the weathering agents may even reach the interior of quartz grains and cause decomposition of the carbonaceaous inclusions (Fig. 16). As a result of the leaching processes microfossils remain preserved as microcavities with their cell morphology as an imprint in the wall. Fossil preservations of that kind are most common in paleontology and suitable objects for morphological studies, but unsuited for chemical investigations. The organic residue present in such cavities may consist of heterogenous components, indigenous matter and contaminants.

It is even possible, that organic particles hermetically sealed in minerals are younger than the hosting sediment and thus contaminants of a more recent time. A sample of chert from the ca. 3500 my old Warrawoona group of Australia has been shown to contain carbonaceous microstructures (Schopf & Walter, 1983) [7]. But critical examination of the thin sections indicates that the specimens are contaminants sealed in a secondary chert that fills a rock fissure and is, therefore, younger than the surrounding sediment (Buick, 1984) [66]. Styliolites and other cracks may form at different times in postdepositional history and may be sufficiently open to allow entry of contaminant microorganisms that filter in from exposed surfaces or are introduced by vadose waters. They may later be sealed by the crystallization of calcite and/or silica (Cloud & Morrison, 1979) [63].

There is yet another problem to be considered. It is now well documented that buried sedimentary rocks, even after compaction, can be permeable and therefore contaminated with younger organic matter introduced by groundwater. The latter problems imply that the identification of biochemical fossils in solvent-soluble substances, is not the optimum approach. An experiment employed by Smith et al. (1970) [67] has demonstrated that the extractable hydrocarbons in certain fossiliferous Precambrian cherts were located primarily at grain boundaries and at other readily accessible sites, rather than being included within the silica crystals. Consequently, it is safer to analyze the insoluble kerogen, rather than solvent extracts of rocks. But Oehler (1977) [68] has shown, using ^{14}C-labelled organic compounds, that even kerogens can become irreversibly contaminated by younger organic compounds.

What safeguards should we take against such postdepositional contamination? The most effective means by which syngenicity can be demonstrated is the optical microscopic analysis of petrographic thin sections (Cloud & Morrison, 1979) [63]. One method utilizes fluorescent microscopy. Most recent and fossil liptinites display fluorescence when irradiated by ultraviolet or blue light. But in the transition from a medium to low volatile rank the intensity of the fluorescence diminishes and finally disappears. This stage can be correleated to the death line for oil in oil source rocks (Teichmüller & Durand, 1983 [69], Van Gijzel, 1982 [70]). Most Precambrian organic-rich sediments have crossed this threshold of intense hydrocarbon generation. As a result they show no fluorescence or a very weak one at best (Fig. 17). Consequently, if notable organofluorescence appears in an ancient rock sample it may contain recent contaminants. In this way the technique can help to decide as to whether a sample is contaminated or not. It should be noted, however, that certain minerals fluoresce quite strongly, e.g. the carbonates. In contrast to most fluorescing organic matter, minerals alter relatively little during irradiation (Teichmüller & Wolf, 1977) [71].

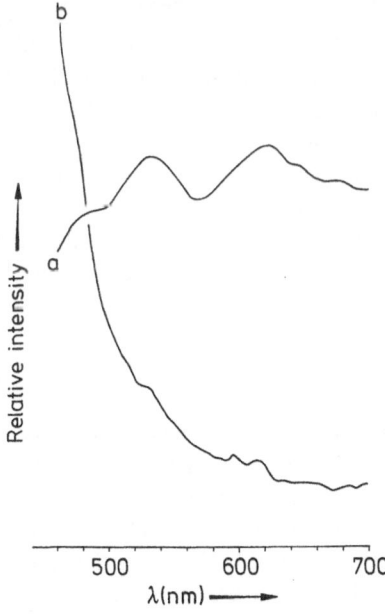

Fig. 17a and b. Spectral distribution of fluorescence energy in the visible region. **a)** Alginite in Permian Boghead coal, **b)** *Huronispora* in Gunflint chert. Irradiation at 365 nm, band width 10 nm, field of measurement 5 μm diam. Courtesy of Zeiss Laboratories

VII Inorganic Chemofossils

This chapter attempts to assess the significance of inorganic chemistry in biological evolutionary processes of the past. Emphasis is placed on the mineral grains and minute mineral aggregates of biological origin that occur associated with the carbonaceous matter in quartz crystals of sedimentary environments. In order to evaluate the importance of the biomineralization processes it is necessary to be able to distinguish biogenic from nonbiologically formed minerals. Certain biogenic minerals usually adopt characteristic shapes defined by the geometry of their organic matrix frameworks. They are often unique in crystal habit, but also in trace element and isotopic composition and occassionally even in mineral species, and this may still be recognized even after their dispersal among the sediments. Striking examples are occurrences of needle-shaped fluorite crystals and magnetite crystals, forming slightly rounded cubes and rectangles, hexagonal prisms with flat ends or teardrop shapes all of which are formed biologically (Kirschvink & Chang, 1984) [72]. Recent observations of mineralized deposits in prokaryotic cells indicate clearly that even bacteria are capable of forming minerals with the elaboration of organic matrices by a "matrix-mediated" process. This could provide a means to improve our understanding of the evolution of biomineralization during the Precambrian (Lowenstam, 1981, 1984) [73, 74].

A more primitive kind of mineral production exemplified by many bacteria as well as various green and brown algae, is characterized by mineral formation without matrix-mediated processes. These processes result in minerals having crystal habits similar to those produced by precipitation from inorganic solutions. Consequently, it is impossible to distinguish products in which microorganisms are involved from those that are purely chemical. As of now, most of the known biominerals still lack physical and biochemical characterization. Data to fill these gaps are desperately needed to establish the beginning of mineral forming processes by life in the Precambrian record.

There is some hope, since certain sediments are thought to be produced only by biological activities. Obviously many of the involved reactions, if not mediated by biocatalysts would require the addition of too much time and heat. In these cases it is plausible to conclude that bacteria had provided the necessary catalysts. Especially, if a specific inorganic formation is found to be associated with organic chemofossils and microfossils, it is very likely a result of biological activities. This is exemplified by stromatolitic formations. When such deposits are found, one can deduce the nature of the biological activities from the characteristics of the mineral formation.

VII.A Metals

Chemical evolutionary processes involving carbon, hydrogen, nitrogen and oxygen have been studied intensively and extensively, but the other essential elements have been neglected in the studies of chemical and biological evolution. Study on the accumulation and ratios of polyvalent metals in fossil matter may help to reveal the bacterial groups to which they belong and their habitats. Two of the most important and interesting inorganic elements in this respect are iron and copper.

When one considers the concentrations in different types of organisms some interesting statements can be made.

Iron is concentrated most by cyanobacteria followed closely by phytoplankton (Jones et al., 1978) [75]. Copper is concentrated most by phytoplankton and next by cyanobacteria. Primitive photosynthesizers such as the cyanobacteria are especially rich in non-heme iron, which is involved in the reduction of CO_2, molecular nitrogen and many other substances. It has been speculated that during the evolution of the plant kingdom, the ratio of iron to other polyvalent metals decreased because the latter became more and more involved in metabolism, chiefly in oxidation reactions in the cells (Ochiai, 1983) [76]. Therefore, cyanobacteria contain much more iron than other plants. It has been also concluded from analyses of individual fossils that the evolution of different algal groupings in the Precambrian was accompanied by a decrease in the iron content and simultaneous enrichment in copper and others (Udel'nova et al., 1981) [77]. Copper has been interpreted to be a marker element of the younger Proterozoic as far as its biological association is concerned. Thus, the two elements iron and copper cover the important period of the Earth's history, between 3.8 — 1.5 and 1.5 — 0.6 Ga resp. (Ochiai, 1983) [76].

A wellknown example of the younger period is the cupriferous Nonesuch shale deposit at White Pine, Michigan. Liquid oil was found there trapped as primary fluid inclusions in calcite crystals which are dated at about 1050 MY. According to the observations the organic materials must have played a significant role in controlling the copper mineralization of the sediment (Kelly & Nishioka, 1985) [78]. Analyses show that the oil contains the full spectrum of hydrocarbons typical of unbiodegraded crude oil. Of the C_{15}–C_{20} acyclic isoprenoids, pristane (C_{19}) is preferentially generated during early catagenesis (Fig. 18). Usually pristane/ phytane values rarely exceed the amount 2 in mature and postmature sediments of Precambrian age (McKirdy & Hahn, 1982) [5].

Most copper enzymes and proteins are found only in eukaryotes, but a few copper proteins, such as azurin and plastocyanin are also present in certain prokaryots

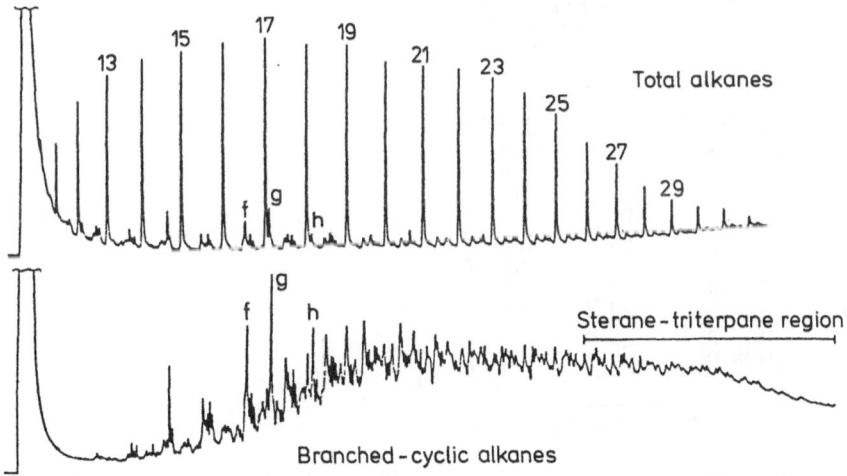

Fig. 18. Gas-chromatograms of C_{12+} total and branched/cyclic alkane fractions of the Nonesuch Shale oil seep, Michigan USA (after Hoering 1976 [79], from McKirdy & Hahn, 1982) [5]

that are aerobic. If the data on copper-mobilization are correct, the cyanobacteria that had developed before the main copper-mobilization (which started about
1,8 Ga ago) should not contain plastocyanin. That is, cyanobacteria of the earlier
type, may not contain plastocyanin, whereas some of the more developed
filamentous cyanobacteria contain it.

Another more primitive mode of biological metal accumulation is the capture of
metals in cell wall and sheath. Certain bacteria can immobilize large quantities of
metal at the cell surface by forming insoluble metal complexes. Anionic ligands
responsible for cation adsorption include phosphoryl, carboxyl, sulphydryl, hydroxyl
groups and others. Subsequently after burial of the cell, the metals may react with
hydrogen sulfide produced by sulfate reducing bacteria and form metal sulfides.
Microprobe analyses have shown that Archean microfossils often contain much copper
in their sheaths (Figs. 19, 30) and in the sulfide facies iron-formations we do find
appreciable amounts of copper sulfides which are clearly cogenetic with the other
facies of the iron-formations.

In spite of the overwhelming evidence that is available for the accumulatory
potential in living systems, the biological involvement in the genesis of fossil
stratified ores has only been recognized in a few instances. In particular the role of the
organic materials in promoting or inhibiting metal adsorption needs to be considered

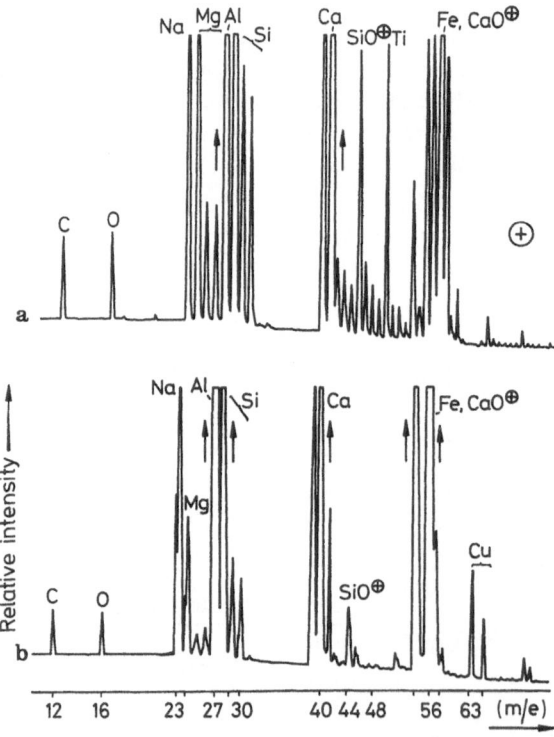

Fig. 19a and b. Laser mass spectra (positive ions) of spherical organic microstructures included
a) in Gunflint chert (*Huronispora*), **b)** in Isua quartzite (*Isuasphaera*). Wall of the spheres is
focussed. Field of measurement 1 μm diameter (Pflug, 1982) [16] (compare Figs. **6a–f).**

in greater detail. The specific effect of this organic matter and complexing ligands on metal adsorption in general are largely unknown.

VII.B Carbonates

Almost all the calcium carbonate that is formed both on the land surface and in the oceans today is produced by living organisms, and it is likely that most of the $CaCO_3$-reservoir of the crust is derived from biologically produced limestone (Monty in Westbroek, 1983) [80]. Carbonate minerals are also by far the most widely utilized bioinorganic constituents.

The literature indicates that consideration must be given to different processes of biological carbonate deposition:

(1) Skeletal structures with matrix development. Cyanobacterium species of the Genus *Geitleria* are known to form a thick mineral deposit on their surface, which is composed of successive layers of elongated needle-shaped calcite crystals (Lowenstam, 1981) [73].

(2) Carbonate encrustations resulting from photosynthetic activity of bacteria and/or algae. This process changes the environmental conditions by reducing CO_2 to organic carbon compounds and thus shifting the solubility equilibrium toward inorganic precipitation of carbonates. In some cases, these are deposited irregularly on cell surfaces.

(3) Carbonate precipitation and lithification caused by the metabolic activities of bacteria dwelling within the environment. These activities bring about carbonate precipitation, liberation of nitrogen and phosphorus. Among the most important geologically active bacteria are sulfate reducers. These bacteria are involved in the formation of H_2S, calcium carbonate, native sulfur and pyrite, and in the reduction of the sulfate ion in water. Below the oxidation zone of the sediment, sulfate-reducing bacteria break down calcium sulfate; as a result, calcite forms. The mechanism for this biochemical reaction depends on the ability of the bacteria to extract oxygen from sulfates such as gypsum. With this oxygen they are able to oxidize organic matter and thus produce energy. When the HCO_3^- produced in this bacterial oxidation reacts with Ca^{2+} ions from gypsum or organic sulfate, calcite forms (Lein, 1978 [81], Friedman, 1978) [82]. The nonfossiliferous condition of certain carbonate sediments in modern seas and in some ancient limestones has been attributed to this biochemical reaction.

(4) Bacterial decomposition in seawater results in a large increase in the concentrations of dissolved bicarbonate, carbonate and ammonia (plus volatile amines). Accompanying this may be a local rise in p_H and the precipitation of Ca^{2+} ion from solution, to form fatty acid salts or soaps with 14 to 18 carbon atoms. The precipitate called adipocere often replaces and takes the shape of the decomposing body and forms concretions retaining much of the external morphology of the organism. During the subsequent breakdown of adipocere it is probable that calcium will form $CaCO_3$ from isotopically light carbon (Berner, 1968 [83], Kazmierćzak, 1979) [84].

The carbonate sediments resulting from the above noted processes are extremely rich in different particle forms such as onkoids, spherulites, flakes, filaments, tubes, chips, sandwiched structures and others (see also Krumbein, 1975 [50]). Certain calcite nodules of cylindrical or subcylindrial shape have been interpreted as zooplankton fecal pellets. Such products have been suggested as the main source of carbonate grains in some Paleozoic sediments (Porter & Robbins, 1981) [85]. Similar particles together with carbonaceous matter have been also reported from Proterozoan sediments (LaBerghe et al. 1984) [86].

VII.C Sulfur Compounds

Sulfur is a key element of life constituting, on average, between 0.5 and 1.5‰ (dry weight) of plant and animal matter. It occurs mainly in proteins that typically display a C/S ratio of about 50/1 (Schidlowski et al., (1983) [87]. Studies on organic sulfur in peats, coals and inorganic soils have led to the characterization of two forms of organic sulfur: (1) a carbon-sulfur linkage such as that found in sulfur-containing amino acids (such as cysteine) and heterocyclics (such as thiophenes) and (2) a carbon-oxygen-sulfur linkage (referred to as "ester sulfate") represented by chondroitin sulfate and phenolic sulfates. When a sediment is buried and becomes increasingly anaerobic, sulfate becomes an important source of oxygen for microbial metabolism. Hydrogen sulfide from the reduction of sulfate can then react with organic matter to form organic sulfur linkages. Bacterial sulfate reduction is associated with isotope effects of varying magnitudes (Casagrande & Siefert, 1977) [88]. More intensive processes of sulfate reduction favor the rapid transformation of the sulfide compounds to the end product, pyrite. The transitional products of this reaction are apparently various metastable iron monosulfides which transform to pyrite during early diagnesis. The processes are controlled by the reactivity of the involved iron compounds. The most reactive fraction is made up of fine grained hydrous ferric oxides (Berner, 1984) [89].

It has been shown by mineralogical, chemical and X-ray-diffraction analyses that the major part of reduced sulfur occurs in the form of pyrite in ancient sediments (Lein, 1978) [81]. It has been also established that pyrite may form rapidly in muds of recent sediments. In anoxic bottom waters, pyrite formation can take place before and after burial even during sedimentation (Berner, 1984) [89]. Also the geological occurrence and chemical stability relations indicate that authigenic pyrites can be synsedimentary or diagenetic (Kalliokosky, 1966) [90].

Pyrite occurs in sediments in the form of single crystals, crystal clusters, spheres, framboids or as replacement for organic structures. Miroprobe analysis of pyritic aggregates often show the presence of appreciable carbon, and some coarser carbonaceous matter is visible microscopically. Microcrystals of pyrites are frequently found in organic particles when examined in the TEM. Sometimes a thin bright rim occurs around each crystal, which indicates that it is enclosed within a carbonaceous shell (Oberlin et al., 1980) [19].

Microfossils are sometimes replaced by either a mixture of framboidal and microgranular sulfides, or by microgranular or fibrous sulfides alone. Thus, pyritized fossils may be distinguished by their external morphology, or by their content of framboidal pyrite. The restriction of framboids to fossils suggests that these forms owe their origin to the presence of organic substances, which may have served as a component essential for the formation of the colloid. Most often pyrite is formed in situ from the organic matter. X-ray photos of cephalopods from Devonian shales proved soft parts to be still present in the shape of delicate pyrite veils (Stürmer, 1984) [91].

In these cases, pyritization of microstructures must have occured before compaction of the sediments. Once the petrifying iron sulfide was deposited, it probably did not undergo further alteration (Fig. 20a).

The occurrence of bacteria-like structures inside of large, well-formed crystals also may deserve consideration. Small inclusions, distinctly different from fluid

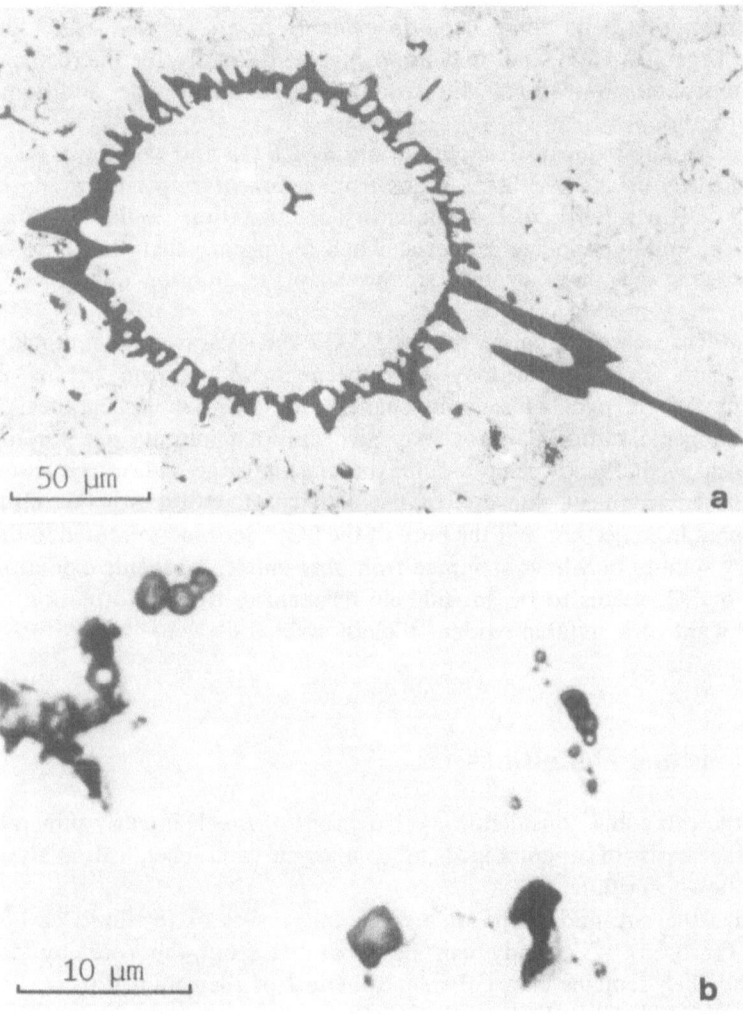

Fig. 20a and b. a) Radiolarian test enclosed in pyrite from the Devonian shale of Meggen (FRG). Silica of the test is replaced by sphalerite (ZnS); **b)** Marcasite containing budding structures interpreted as pseudomorphs after *Pyrodictium*. From sulfide formation of the East Pacific rise (Tufar et al., 1984) [92]

inclusions were found in marcasite and pyrite deposited around submarine hydro-thermal vents. Bud-like structures with some major buds and minor buds at the ends of filament-like structures resemble bacteria of the *Pyrodictium*-type (Tufar, 1984) [92] (Fig. 20b).

Ramdohr (1953 [93]) has found inclusions within sulfide minerals of the Mid-Devonian Rammelsberg ore deposit in Lower Saxony, which he attributes to pyritized bacteria. Similarly, in the Kupferschiefer formation spherical aggregates of various sulfides, among them pyrite, were regarded as mineralized bacteria (Harder, 1919) [94]. Electron micrographs of the oxidized surface of a Middle Pennsylvanian pyrite specimen show fossil bacteria with good resolution of morpho-logic detail (Schopf et al., 1965) [95]. Similar structures have been repeatedly described from pyrites of Precambrian sediments of various ages (Love & Zimmermann, 1961 [96], Muir, 1978) [45]. So there is no reason to doubt the presence of abundant bacteria in ancient pyrites, and some bacteria evidently are preserved as fossils. Nevertheless, the organisms are small and impose great difficulty for microscopic observation of morphologic features. Electron micrographs provide additional morphologic information.

The existence of sulfate-reducing bacteria as old as 2.8 Ga and possibly 3.2 Ga has been established by means of $^{34}S/^{32}S$ ratios from sedimentary pyrites found in Precambrian rocks. The values show dissimilatory fractionation similar to those produced by extant sulfate-reducing bacteria. Thus it appears that this type of mineralization process, may have evolved in the early Precambrian (Schidlowski et al., 1983) [87].

The widespread occurence of barite in the 3.5 Ga old Warawoona sediments of Australia indicates that sulfate may have been present in the oceans in abundance already 3.5 Ga ago. This could suggest that the first appearance of presumably bacteriogenic sulfide patterns may give only a minimum age for the underlying biological event. Since we may assume that the first large-scale introduction of sulfate into the environment was due to the activity of photosynthetic sulfur bacteria it is reasonable conjecture that the bulk of the SO_4^{2-} ions incorporated in the oldest sedimentary sulfates may have stemmed from this source. Inorganic oxidation of volcanic H_2S or SO_2 seems to be an unlikely mechanism for the formation of these sulfates for want of a suitable oxidant (Schidlowski, 1984a, b) [97, 2].

VIII Conclusions and Definitions

It appears from the preceding considerations that proof of fossils in early minerals is possible on the basis of a combined morphological and chemical analysis comprising the following criteria.

(1) The find has the size and shape of a cell and consists of fossilized carbo-naceous matter (kerogen). The body can be separated from the rock by de-mineralization and then remains as a coherent spherical or filamental skin.

(2) Organic compounds identified in the microstructure are typical alteration products of carbohydrates, hydrocarbons, porphyrins and related plant materials. Their metamorphic condition is consistent with that of the hosting mineral.

(3) The contained carbon and sulfur resemble that of biological materials in their isotopic composition.

(4) The microstructure in situ is often externally encrusted and internally filled with carbonates, pyrite or other mineral matter of apparently biogenic origin.

(5) The find is associated with other specimens of the same kind and together with non-structured carbonaceous debris of similar chemical composition. The assemblages are arranged along bedding planes or other primary patterns of the sediment which is a chert, stromatolite, banded iron formation, shale or related rock.

(6) Well preserved specimens show structural details in their cell wall and sheath. Others are present in typical stages of reproduction or growth. Individuals are united in clusters, chains, ramifications or other definable arrangements.

Using the above crieria, biological entities can be distinguished from similar looking objects of non-biological origin in most cases. This will be demonstrated with representative examples in the following chapter.

IX Examples

IX.A Gunflint

The oldest generally accepted microbiota occur in the ~2.0 Ga old Gunflint iron formation in Ontario, Canada. There are reports that morphologically simpler microfossils are present in various older rocks, back to 3.8 Ga ago. Thus, it would be beneficial to gain paleobiochemical data supportive of these very old and simple microstructures. The Gunflint is obviously a suitable model on which the chemical and structural features of ancient life can be demonstrated.

Several types of organic microstructures, filaments, spheres and others, ranging from a few micrometres to millimetres in diameter, are recognized in the Gunflint thin Sections. They commonly occur embedded in a chert matrix and are often contoured by finely grained carbonate. Their interior is usually filled with microcrystalline quartz identical to the surrounding matrix. In many spheres, the center contains dolomite or siderite aggregates more rarely pyrite inclusions (compare Fig.s 6, 21a, b). Grains of pyrite are often also dispersed along the internal surfaces of the wall. Other pyrite matter occurs in a variety of forms, as a replacement which may replicate filaments, microspheres, ooliths, and other structures. Minute pyrite grains less than 1 μm in diameter tend to outline the organic structures producing an effect similar to stippling (compare Fig. 21d). In a few instances the organic filaments and spherical bodies are composed of solid pyrite. Individual filaments of solid pyrite have been observed to pass laterally and continuously into carbonaceous films containing scattered pyrite grains less than 1 μm in diameter (Barghoorn & Tyler, 1965 [98]), Lougheed, 1983 [99]), Krumbein et al., 1984) [100]).

The spherical structures (Huronispores) are commonly clear chert surrounded by an amorphous, light yellow organic substance (Fig. 6a–c). This material is also observed to preserve, or represent spherical bodies within disseminated carbonate rhombs included in the chert (compare Fig. 34). Microprobe analysis shows that the

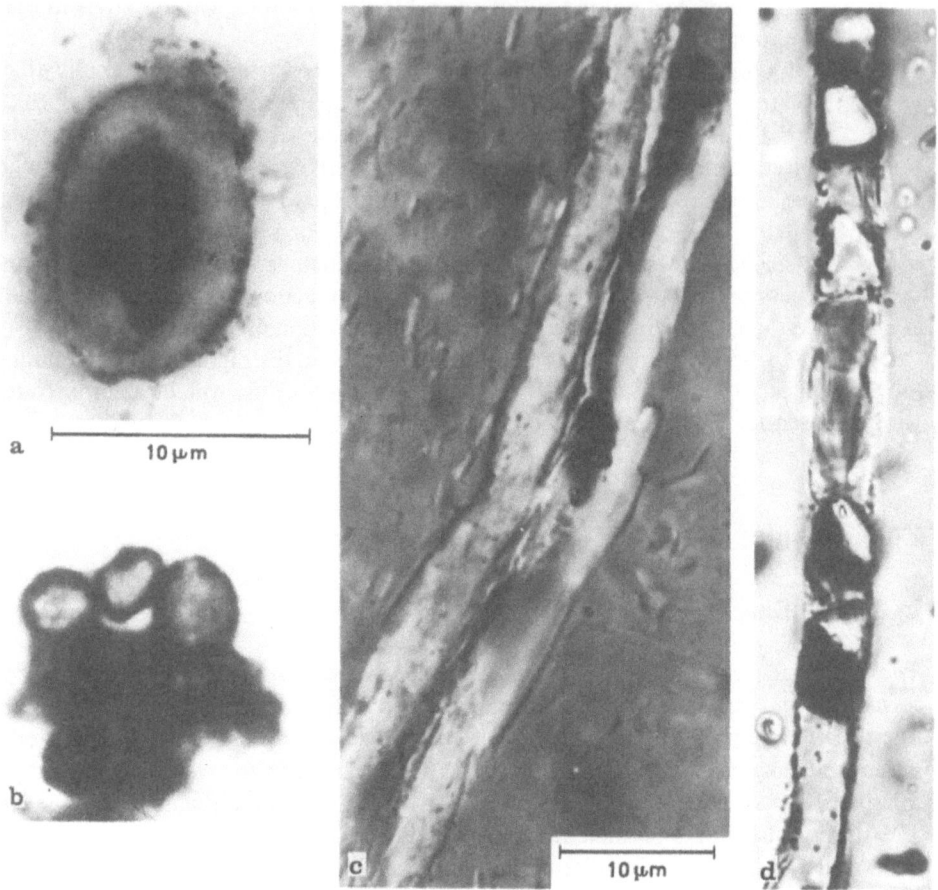

Fig. 21 a–d. *Huronispora* outlined by minute grains of pyrite and filled interriorely by black pyrite, **b)** Cluster of the *Bavlinella* type (compare Fig. 25), **c)** Filamental structure preserved with its carbonate sheath (bright), **d)** Filamental structure contoured by pyrite grains (black). Figs. **a) b)** from Gunflint chert, Figs. **c) d)** from Swartkoppie (Onverwacht) chert. Bar for Fig. **a) b)** see Fig. **b)** bar for Figs. **c), d)** see left of Fig. **c** (Pflug 1985) [18]

yellow substance is a dilution of kerogen in quartz sometimes intermixed with carbonates or iron oxides. It is evident that the structures are best represented by these altered remnants of the original organic matter.

The typical infrared spectrum shown in Fig. 22 has been obtained from a colony of Huronispores demineralized before measurement. The wide asymmetric band centered at 3400 cm^{-1} is related to OH groups (phenolic, alcoholic, carboxylic OH). The group of signals falling between 2850 and 3050 cm^{-1} may be assigned to C—H vibrations. The weak shoulder on the high frequency side of the bands is probably due to the 3030 cm^{-1} band representing aromatic C—H stretching. The bands comprizing the signal (Fig. 22b) are the stretching vibrations of the saturated methyl- and methylene-groups. The strongest methylene band appears at 2920 cm^{-1}, and another methylene frequency occurs at 2855 cm^{-1}. The weak depression between

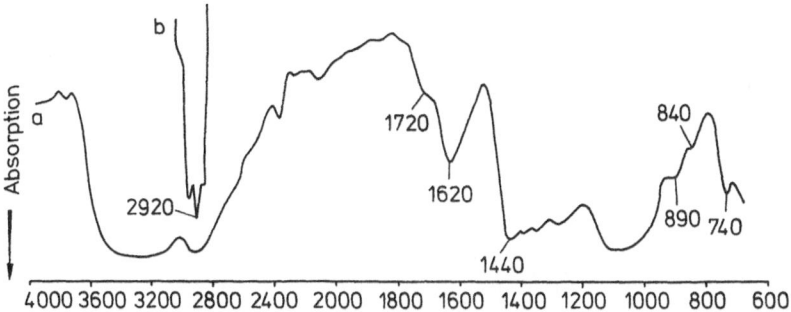

Fig. 22a and b. Infrared spectrum obtained from a cluster of *Huronispora* (Gunflint chert) in demineralized condition (compare Figs. **6a–c**); **b)** High sensitivy recording of CH stretching region. Field of measurement 25 μm diam. (Pflug, 1985) [18]

1730 and 1700 cm^{-1} and part of the band at 1620 cm^{-1} are attributed to double bonded oxygen. Absorption in the 1735 cm^{-1} region is characteristic of open chain carbonyle groups, whereas the portion around 1700 cm^{-1} may be due to aryl ketones. The bands at lower frequencies represent bending structures. The valley at 1440 cm^{-1} is due to either asymetric $C—CH_3$ or methylene. A doublet in the range 1380–1360 cm^{-1} may indicate the terminal carbon atoms of microbial fatty acids. Absorption between 1400 to 1040 cm^{-1} includes $C—O$ stretching and OH bending. The flat depression at 890 cm^{-1} indicates isolated hydrogen, the one at 840 cm^{-1} two adjacent hydrogens and that at 740 cm^{-1} four adjacent hydrogens. The spectrum as a whole is comparable to known spectra of kerogens (Robin & Rouxhet, 1976) [101] and asphaltenes (Yen & Erdmann, 1962 [102], King et al., 1963) [103].

Raman spectra from the cell wall and sheath of Huronispores are characterized by peaks at 1610, 1350 and 2700 cm^{-1} which are assignable to $C=C$ double bonds (Fig. 23). The type of spectrum will be considered in detail on p. 40ff.

The organic fraction of the darker and more carbonaceous samples of the Gunflint chert ranges from 0.2 to 0.6% (dry weight), the average for the darker samples beeing 0.37%. The organic residue, after demineralization in hydrogen fluoride, yielded small amounts of soluble compounds when extracted with benzol and methanol. The extracts fluoresce strongly in ultraviolet light. Fractions eluated by the heptane and carbon tetrachloride fractions are presumably all alkane hydrocarbons, and the benzene eluate is presumably made up of aromatic hydrocarbons, probably consisting primarily of phenanthrenes, as shown by ultraviolet absorption. The methanol nonhydrocarbon eluate was subjected to infrared absorption studies. The infrared absorption spectrum indicates the presence of carbonyl bonding and alcohol and ester linkages (Barghoorn & Tyler, 1965) [98]. Since extraction products may contain younger contaminants, the analyses should be considered with caution.

Rapid embedding in silica is particularly favourable for the good preservation of the organic matter, and this was surely the case with the well-preserved microstructures. But there is much indication from the thin section petrography that the Precambrian banded iron-formations were primarily deposited as carbonates (Lougheed, 1983) [99]. The origin of the embedding chert has been a major problem,

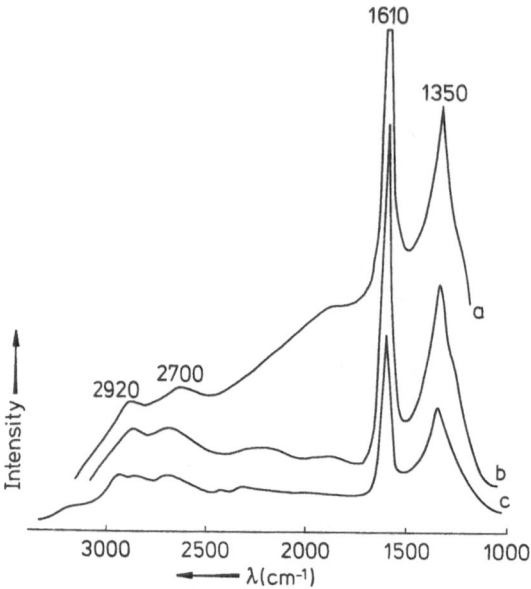

Fig. 23a–c. Raman spectra from cell walls of *Huronispora* included in thin Section of the Gunflint chert before **a)** and after demineralization **b, c)** (compare Figs. **6a–c).** Field of measurement ca. 1 µm diam. (Pflug, 1982, 1985) [16, 18]

but the textural relationships indicate that most, or all, of the chert is a replacement of earlier minerals. The textures in the chert are in every way analogous to textures in recent calcareous sediments, indicating that many iron formations originally were siderite-rich limestones. Although silica-secreting organisms may have existed (Klemm, 1979) [8], the petrographic evidence is that most, or all, of the chert is secondary.

Kazmierćzak [84] distinguishes several stages in the preservational history of the microbiota: living metabolic activity, early post-mortem (preburial) degradation and initial ferro-carbonate precipitation, shallow burial diagenesis resulting in final sideritic permineralization (Fig. 24). Oxygen released from photosynthesizers may have reacted with ferrous iron to form ferric compounds, probably hydroxides, which were concentrated on and within the mucilage surrounding the cells. The coenobia embedded in the anoxygenic bottom zone must have been subjected to intensive bacterial degradation, decarboxilation of amino acids and proteolysis leading to the formation of ammonia and release of carbon dioxide. The presence of ammonia probably caused a considerable increase of p_H, producing dissolution of silica in the intermediate surroundings of the decaying colonies and facilitating, the reaction of CO_2 (as CO_3^{2-} species) with Fe^{2+} to initiate siderite precipitation around the decomposing cells. Probably, the iron of the ferrocarbonate was previously dissolved in water (Fe^{2+}), but could also have been metabolically produced in the mucilage (as Fe^{3+}), and after reduction it was also involved in the reaction with carbon dioxide (CO_3^{2-} species) to form $FeCO_3$.

Sideritic spherical structures of the *Huronispora*-type are abundant in BIF and have been described in a number of papers (LaBerge, 1967 [104], Gross, 1972) [105].

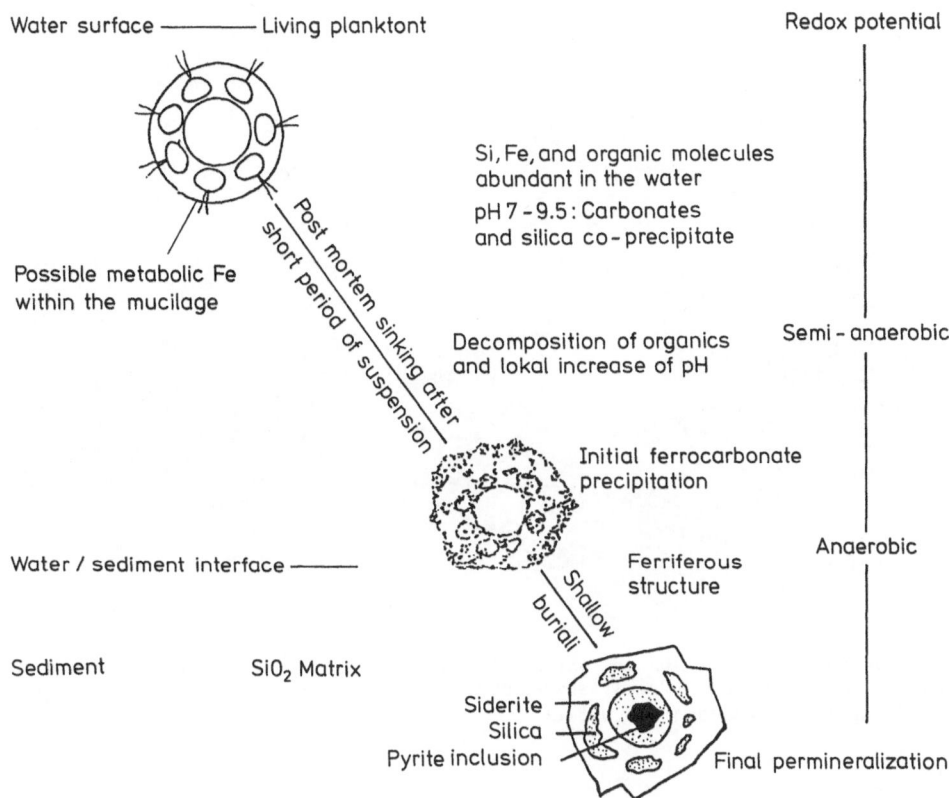

Water surface ——————— Living planktont

Redox potential

Si, Fe, and organic molecules
abundant in the water
pH 7 - 9.5 : Carbonates
and silica co-precipitate

Possible metabolic Fe
within the mucilage

Post mortem sinking after short period of suspension

Decomposition of organics
and lokal increase of pH

Semi - anaerobic

Initial ferrocarbonate
precipitation

Water / sediment interface ————

Ferriferous
structure

Anaerobic

Shallow buriali

Sediment   SiO$_2$ Matrix

Siderite
Silica
Pyrite inclusion

Final permineralization

Fig. 24. Suggested stages of post-mortem degradation and permineralization of the Gunflint spherical
microstructure *Eosphaera* (Kazmierčzak, 1979) [84]

They are especially common in the laminated subtidal siderite-chert facies that
contains much organic matter and pyrite.

IX.B Bulawaya and Witwatersrand

The Bulawayan stromatolite of Simbabwe/South Africa is about 2.8 Ga old and
consists mainly of carbonates and chert. Petrological and electron microprobe
analysis did not reveal any sign of even moderate metamorphisms. Spherical
microstructures are found embedded inside the rock in thin sections (L. A. Nagy &
Zumberge, 1976) [106]. The particles were individually analyzed by the electron
microprobe and found to contain ca. 2 ± 0.5% organic C mostly mineralized
with dolomite. The fact that organic carbon is present in these microstructures
was also confirmed by laser mass analyses which shows a series of polymer
breakdown products similar to those obtained from the ·Gunflint microfossils
(Figs. 25, 26).

The interesting compounds furaldehyde and benzonitrile were detected by vacuum
pyrolysis-gc-ms. These products were interpreted respectively as remnants of biological

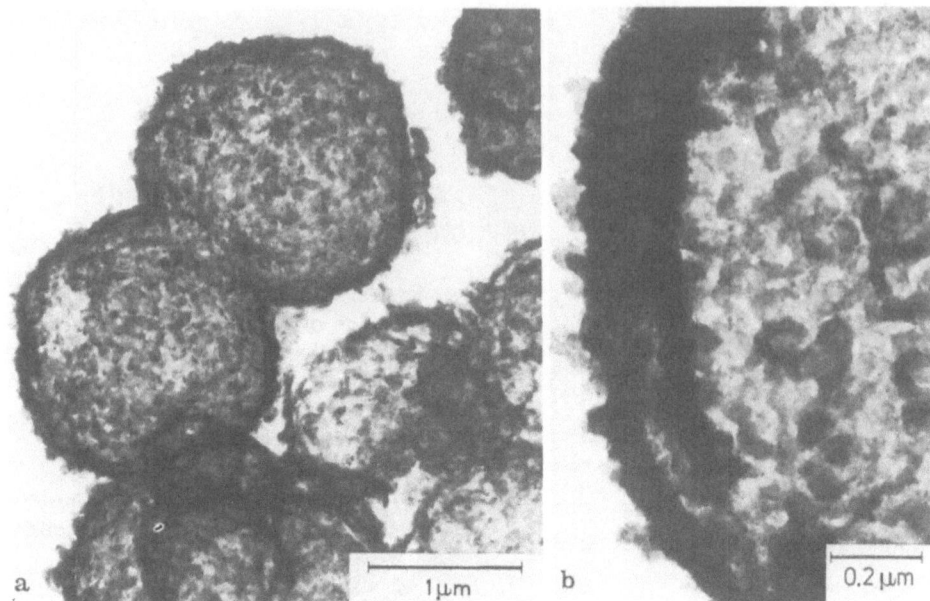

Fig. 25a and b. Transmission electron micrographs of organic spheres (*Bavlinella*-group, cf. Cyanobacteria) from the Bulawaya stromatolite, Zimbabwe. Fig. **b)** is a portion of the cell wall in higher magnification (Pflug, 1984c) [108]

carbohydrates, proteins, and fatty acids (Sklarew & Nagy, 1979) [107] (Fig. 27). Additionally acyclic C_{14}–C_{20} regular isoprenoids, notably pristane and phytane, have been identified (McKirdy & Hahn, 1983) [5], (Fig. 28). Potential biochemical precursors include the phytyl (C_{20}) alcohol side chain of chlorophyll and membrane ethers of archaebacteria. The microstructures found in situ and in demineralized condition (Fig. 25) resemble the *Bavlinella*-type (Fig. 21b) which is commonly interpreted to belong to planktonic endosporulating cyanobacteria (Hofmann, 1984) [110]. Their great abbundance in the sediment is possibly due to explosive blooms in a stressed habitat. Archaebacteria, the methanogens in particular, could also have contributed to the hydrocarbon fraction.

The ~2600 MY old Witwatersrand gold/uraninite ore deposits in South Africa locally contain carbonaceous matter concentrated in small seams. Vacuum pyrolysis revealed a variety of aromatic hydrocarbons and aromatic sulfur compounds but only a few aliphatic components (B. Nagy, 1976) [111]. Apparently, the ionizing radiation arising from the uraninite has altered the original organic matter. The carbon isotope composition of the kerogen points to a biological origin (Hoefs & Schidlowski, 1967) [112]. Structures resembling microorganisms have been repeatedly detected in the matter. The most common form is a mat composed of vertical finger-shaped columns of sub-millimeter size. These consist of a network of filaments which have many features in common with lichens (Hallbauer, 1975) [113]. The surface of the filaments are often encrusted and impregnated by gold, uranium-oxide and silica. It is suggested that these substances were brought into the environment as organic protected colloids. As is known from recent examples, lichens can cause flocculation and precipitation of such colloids (Dexter-Dyer Grosovsky, 1983) [114]. Among the reactional products are organo-uraniferous compounds. Such complexes have been identified in the kerogen of uraniferous sediments from the Precambrian of Canada and Gabon (Rouzaud et al., 1980) [115].

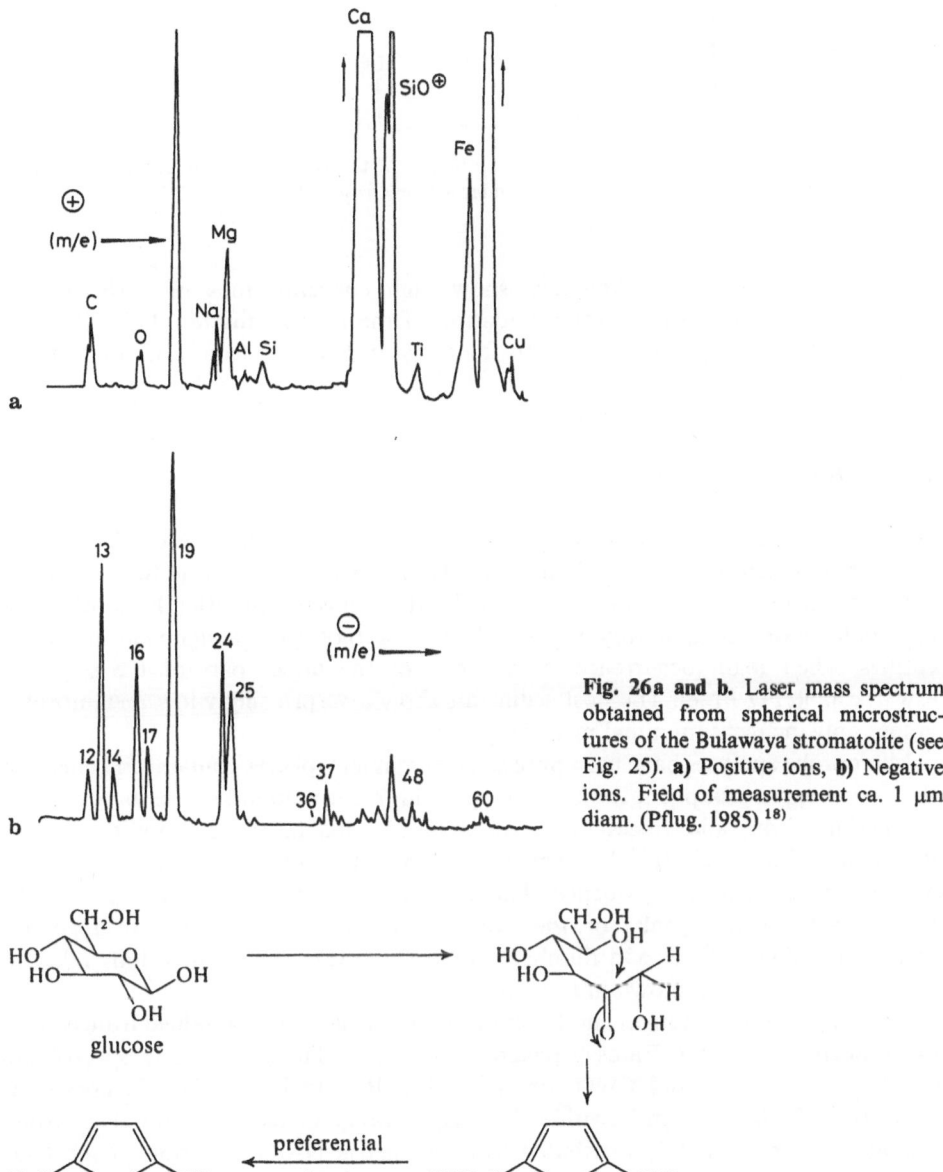

Fig. 26a and b. Laser mass spectrum obtained from spherical microstructures of the Bulawaya stromatolite (see Fig. 25). a) Positive ions, b) Negative ions. Field of measurement ca. 1 μm diam. (Pflug. 1985) [18]

Fig. 27. Suggested reactions yielding 2.5-dimethylfuran from glucose (B. Nagy, 1982) [109]

39

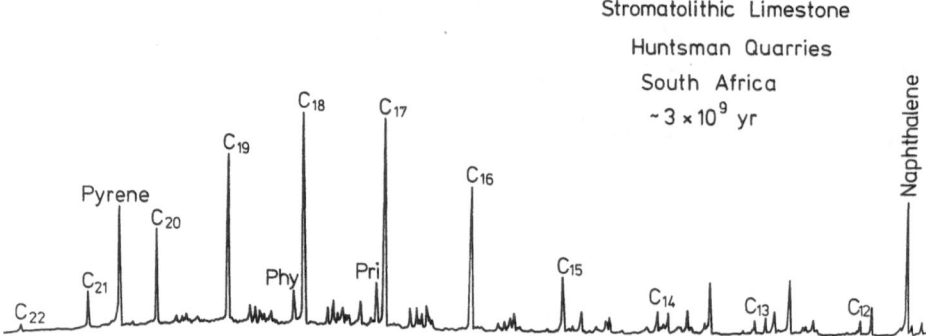

Fig. 28. Gaschromatogramm of volatile organic matter of the Bulawaya stromatolite, Zimbabwe. Pyrene and naphthalene added as internal standards (McKirdy and Hahn, 1982) [5]

The IR spectra of these kerogens show high concentrations of carbonyl and carboxyl groups indicating strong oxidation of the organic matter. This alteration probably results from a reaction with free oxygen delivered from a photosynthetic source.

IX.C Onverwacht and Fig Tree

Organic microstructures have been described repeatedly from the ca. 3.4 Ga old Archean sedimentary rocks of the Swaziland System in eastern South Africa. Most of them are such simple spheroids that, given only the kind of gross morphology we could hardly be confident of a biological assignment for these entities. Their main occurrence is in cherts of the upper part of the sequence (Swartkoppie/Fig Free). The host sediments show a surprisingly low metamorphic grade; only incipient, minimal metamorphism.

Microprobe analyses have been published from microspheres contained in cherts of the Swartkoppie group (Figs. 29, 30). Copper in larger amounts, accessory calcium, magnesium, iron, and traces of nickel were detected in the exterior coating of the bodies (Pflug, 1967) [116]. Carbon, oxygen, and sulfur detected in the wall of the spheres provided evidence that carbonaceous material is present. In the Laser mass spectra, peaks of the even numbered C_nH clusters are regularly higher than those of the odd numbered ones (Fig. 31). The positive ions released from the specimen are shown in Fig. 30.

Interpretable Laser Raman and Laser mass spectra have been obtained from certain other microstructures ("Ramsaysphaera") (Fig. 32). The Laser mass spectra are characterized by CN and CNO ions (Fig. 33). Raman lines (Fig. 32) appear at 1360, 1600, 2720, 2960 cm^{-1} within the organic range of the spectrum. The strong line at 1360 cm^{-1} may be atrributed to a symmetric N—O vibration of the NO_2 group, the weaker line at 1600 cm^{-1} is characteristic of aromatic double bonds C=C. The first overtone of the 1360 cm^{-1} line is observed at 2720 cm^{-1}. The spectrum has the features of a resonant Raman spectrum. It is very often obtained with this type of product in which a large delocalisation of electrons is possible.

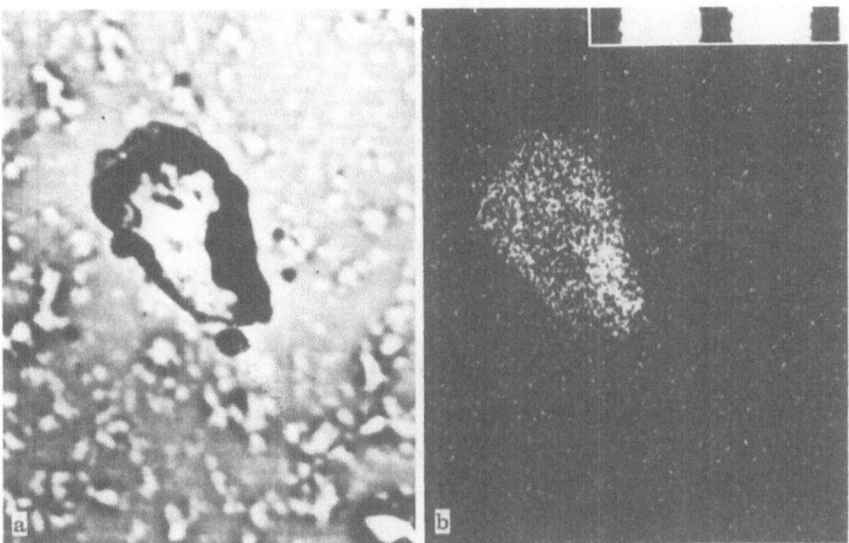

Fig. 29a and b. Spherical organic microstructure (*Archaeosphaeroidites* type) in Swartkoppie chert.
a) Composition scanning immage. The black coating indicates that elements of high atomic number
are present (copper in larger amounts, accessory calcium and iron). **b)** X-ray immage: Sulfur (K alpha).
Scale = 10 μm (Pflug, 1967) [116)]

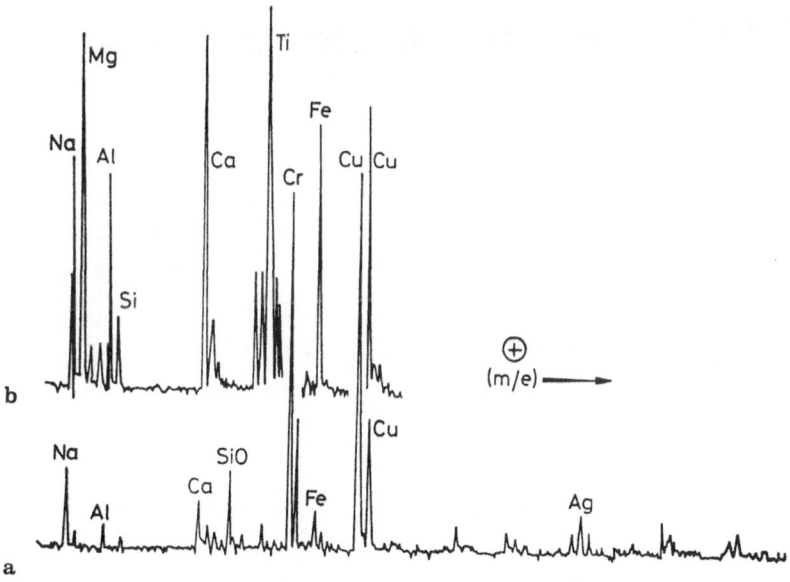

Fig. 30. Two Laser mass spectra (positive ions) from spherical microstructures (*Archaeosphaeroidites*) in Swartkoppie chert. Field of measurement ca. 1 μm diam. (Pflug, 1985) [18)]

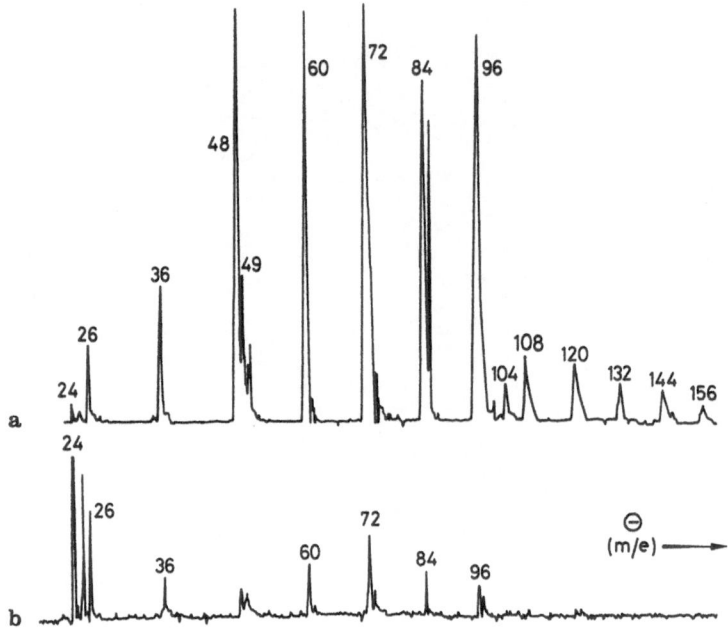

Fig. 31. Laser mass spectra (negative ions) of the objects Fig. 30

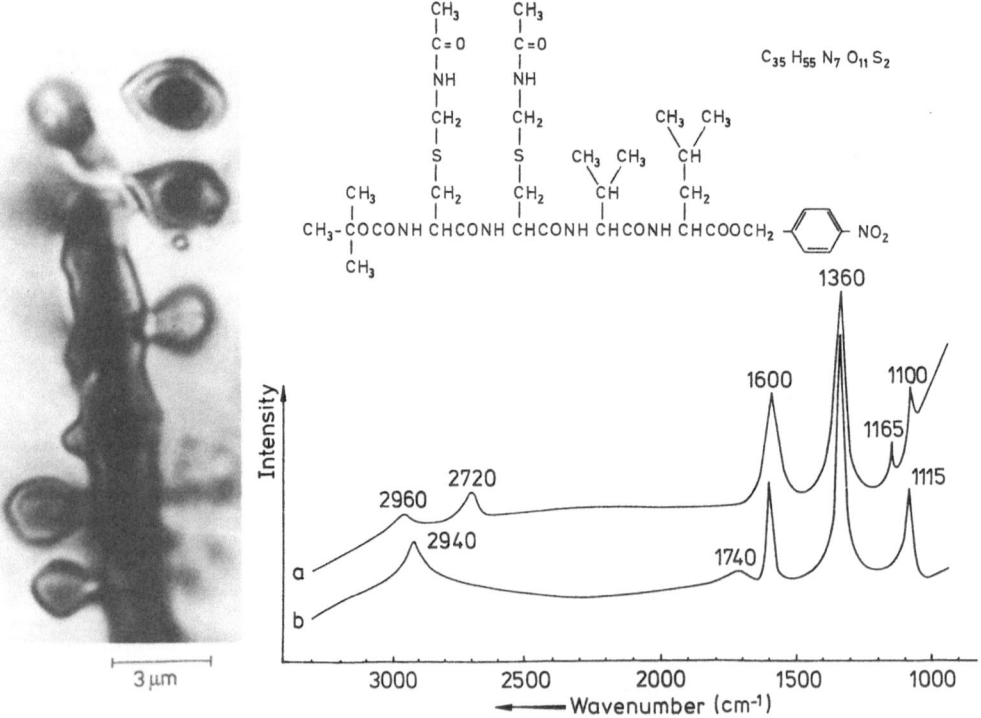

Fig. 32a and b. a) Laser Raman spectrum from a organic microstructure (*Ramsaysphaera*) in thin Section of Swartkoppie chert. Field of measurement ca. 1 μm diam. b) Comparison with similar (but not identical) reference spectrum (Pflug, 1984c) [108]

In this case, the small intensity of the peak at $2960 \, \text{cm}^{-1}$ is also due to this effect, if it is attributable to CH stretching vibrations.

Conclusively, the material is characterized by an N-linked aromate, which, together with additional aliphatic and aromatic structures, and probably under participation of sulfur and phosphorus, composes complex molecules. In the exterior sheath of the microspheres, dolomite is indicated by the lines at 1100, 725 and $300 \, \text{cm}^{-1}$. A weak line at $1445 \, \text{cm}^{-1}$ might belong either to dolomite or to a C—H deformation (Pflug et al., 1979) [117] (Fig. 34). There is scarcely any doubt that the detected organic substances represent decomposed and fossilized remains of cell material.

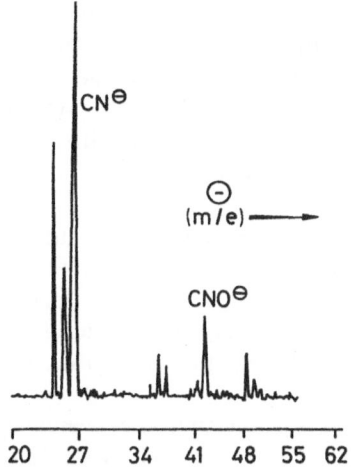

Fig. 33. Laser Mass spectrum (negative ions) showing CN^- and CNO^- clusters in the structural type shown in Fig. 32. Field of measurement ca. 1 μm diam

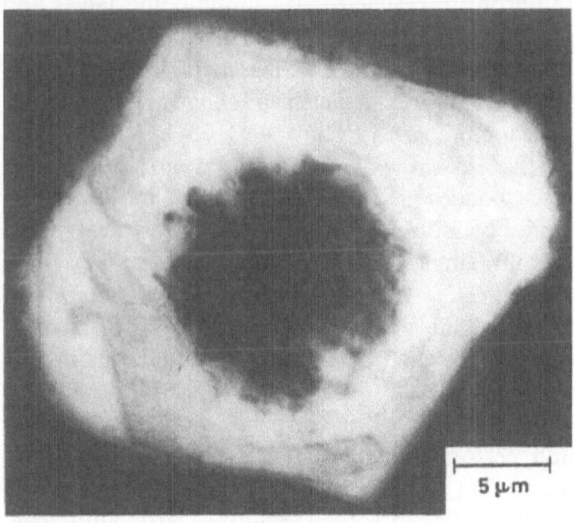

Fig. 34. Spherical organic microstructure included in dolomite crystal of Swartkoppie chert (Pflug, 1985) [18]

Kerogens isolated from the Fig Tree cherts produced very complex mixtures of pyrolysis products, dominated by a series of methyl branched alkenes with each member of the series having 3 carbon atoms more than the previous member. At each carbon number a highly complex mixture of branched alkanes and alkenes plus various substituted aromatic compounds was found. The highly branched structures may have actually incorporated isoprenoids originally present in the Precambrian microorganisms (Philp & Van DeMent, 1983) [6].

IX.D Isua

In the 3.8 Ga old Isua supracrustal belt, West Greenland, a banded iron-formation occurs. The sequence can be subdivided into different facies according to composition and mineralogy, and these facies resemble those of younger Pre-cambrian iron-formations. The carbonate facies in the Isua is mostly finely layered and composed of a magnesium- and manganese-rich siderite locally with thin magnetite-rich bands. Carbonaceous matter, occurring in round aggregates up to 200 μm in size, can amount to more than 5 vol. %.

The silicate facies is of a black finely laminated type, consisting mainly of magnetite with up to 15 vol. % grunerite and up to 5% carbonaceous matter. The facies is found in layers up to a few dm thick. Common to the silicate facies is the scarcity of quartz and the complete absence of carbonates. The sulphide facies of Isua consists of up to 60 vol. % sulphides (pyrite, pyrrhotite) together with grunerite or actinolite and magnetite (Appel, 1980) [118].

Models for the formation of Precambrian sediments suggest that the chemical sediments (such as cherts) of the Isua supracrustal belt have formed as shallow water deposits. This is in agreement with structures locally preserved in the metacherts of the sequence. After deposition, the supracrustals were folded and metamorphosed. Finally, the metamorphism reached lower amphibolite facies and in consequence, most of the primary minerals became recrystallized. As a result all chert now appears as quartzite. But apparently metacherts, magnetite iron formation and quartz carbonate rocks have retained their major element chemistry largely un-altered during metamorphism (Nutman et al., 1984) [119].

Nagy et al. (1975) [120] were the first to show that the Isua quartzite contains microstructures that are evidently carbonaceous and indigenous to the sediment. Spherical entities were found along bedding surfaces and a large percentage of the objects occur enclosed within single quartz crystals, where they may form scattered clusters or bands (Fig. 35a). These particular bands contain little or no actinolite. In typical cases, the spheres consist of a core of clear quartz or poorly defined calc-silicates surrounded by a rim of carbonaceous matter and/or carbonate grains (Fig. 35b).

Several features indicate that the silica spheres are original constituents of the rocks, Isuaspheres have been found perforated by needles of actinolite (Fig. 36). It is obvious from this observation that the spheres must be older than the metamorphic event that produced the needle. In some specimens, large authigenic crystals of siderite or ankerite, enclose the silica spheres (compare Fig. 34). The rhombohedral crystal faces of siderite often extend considerably, making

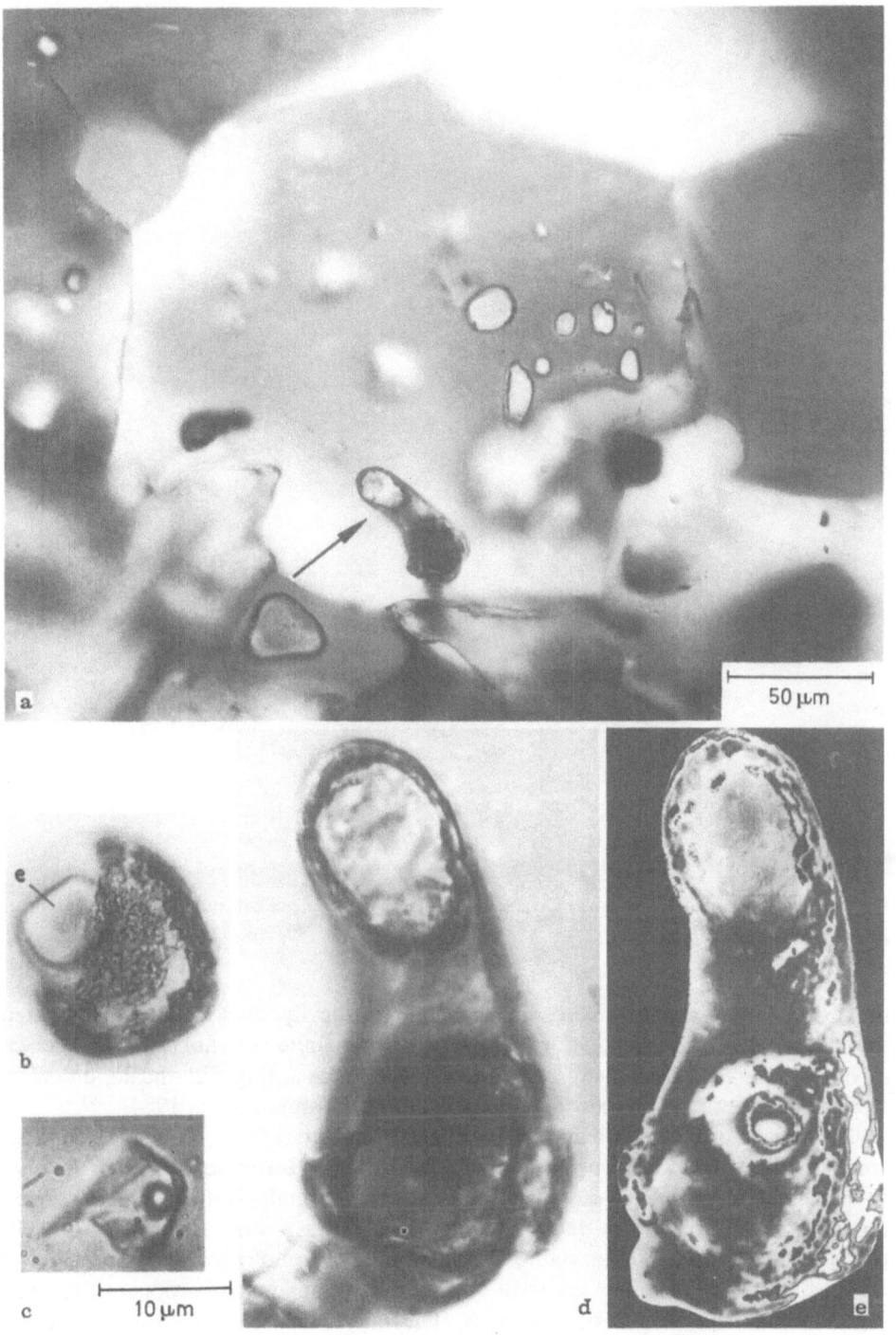

Fig. 35a–e. Isuasphere included in quartz crystal of Isua quartzite (arrow in Fig. **a**), in higher magnification **d**) and equidensity image **e**) (Pflug, 1985) [18]. Fig. **b**) is a sphere with a core of calc-silicate (**e**). Fig. **c**) shows a typical fluid inclusion in the quartzite. Bar for Fig. **a**) see left below the photograph, bar for Figs. **b–e**) see below Fig. **c**). Fig. **c**) is from E. Roedder

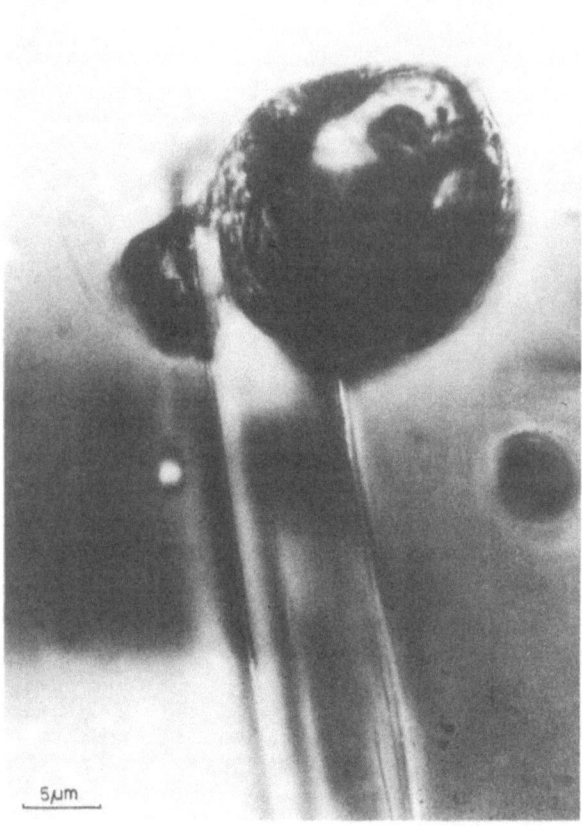

5 µm

Fig. 36. Isuasphere included in Isua quartzite perforated by a needle of actinolite of the metamorphic period (Pflug, 1985) [18]

the external outlines of microspheres characteristically angular. It is suggested that once the calc-silicate formations were dolomite or ankerite and chert mixtures, whereas other chemical sediments were once authigenic silicate, dolomite or ankerite, siderite, iron oxide and chert mixtures (Nutman et al., 1984) [119].

The basic morphological and mineralogical characteristics of the microspheres correspond to those known from other banded iron formations (Fig. 6, p. 10). It is actually impossible to establish any clear-cut morphological difference between the Gunflint-Huronispores (commonly interpreted as cyanobacteria) and well-preserved Isuaspheres. The structures display similar morphological and mineralogical features indicating that they are both of the same origin and the product of similar, degradational and diagenetic processes. The only notable difference is that the Isuaspheres have been subjected to a more intense metamorphism.

In weathered samples of the Isua quartzite, most carbonaceous material has been leached out along with the associated carbonates, leaving spherical cavities in the place of the Isuaspheres. These have been mistaken for fluid inclusions by

Bridgwater et al. (1981) [121] whose erroneous interpretation was subsequently corrected by Roedder (1981) [122], (comp. Figs. 35b, c, d). As can be shown, the cavities form mainly by dissolution of carbonaceous matter and the ferruginous dolomite. So they represent merely an intermediate stage in the destruction of former microspheres (Fig. 16, p. 24).

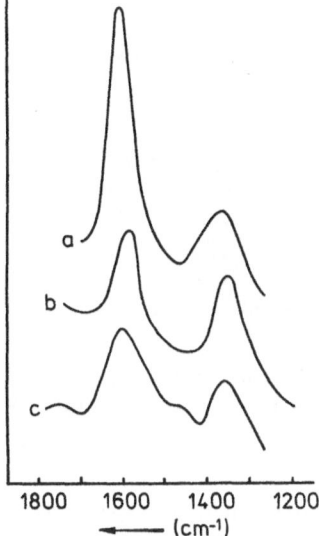

Fig. 37. a. Laser Raman spectra from the wall structure of Isuaspheres (see Figs. 6d–f). Field of measurement ca. 1 μm diam. (Pflug et al., 1979) [117]

According to the Raman spectra and the Laser Mass spectra, the *Isuasphaera* carbonaceous matter present in unweathered samples consists mainly of a high rank kerogen and/or amorphous carbon. It is possible that some graphite is additionally present in the matter (Fig. 37). The Raman spectrum of graphite is characterized by a strong very sharp line at 1580 cm^{-1} which is clearly different from the broader band usually occurring in Isuasphaera spectra at about 1595 cm^{-1}. Another difference is found in the band at 1360 cm^{-1} which appears to be narrower and better separated in the graphite spectrum than in the Isuasphaera spectrum. Thus, graphite can account, for at at best some of the fossil matter. As is known oxygen-rich organic matter such as cellulose produces a carbon which cannot be transformed into graphite. At the highest degree of catagenesis, this material practically turns into a pure carbon. Aromatic CH and free radicals tend to disappear. However, there is no crystal growth (Oberlin et al., 1980) [19].

A peculiar type of microstructures occuring abundantly in the Isua banded iron formation can be made visible under the TEM (Pflug, 1984b) [123]. These have the form of sheathal tubes packed together in bundles or layers (Fig. 38). In places, spherical bodies are enclosed in the tubes. Laser Mass spectra obtained from the particles indicate a high rank carbonaceous matter that is not completely graphitized (Fig. 39). Iron and some copper is regularily associated with the sheaths. Morphologies of a similar kind have been repeatedly reported from other iron formations of the Precambrian (Muir, 1978) [75]. They were usually compared to iron

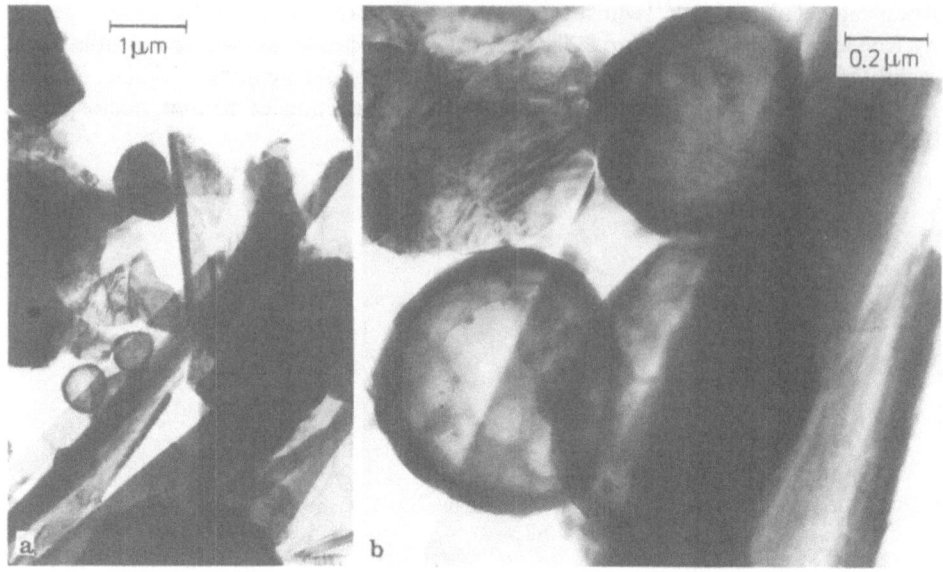

Fig. 38a and b. Microstructures resembling iron bacteria contained in the Isua banded iron formation. Fig. **b)** shows a portion of Fig. **a)** in higher magnification. Electron micrographs from demineralized rock section (Pflug, 1984b) [124]

bacteria of the *Sphaerotilus/Leptothrix* group. A caracteristic of these organisms is the tendency to deposit large amounts of ferric iron in and on their sheaths. It has been repeatedly suggested that iron bacteria and photosynthetic organisms, especially cyanobacteria are involved in the deposition of the BIF (Melnik, 1982) [44].

Bridgwater et al. (1981) [121] have claimed that a biogenic interpretation of the carbonaceous microstructures is inconsistent with the metamorphic history of the Isua region. But Roedder (1981) [122] has pointed out that the presence of dolomite in contact with quartz places constraints on the maximum pressure-temperature combination of the metamorphism. The two minerals react to form diopside plus CO_2 at 2000 bars at 520 °C; each 20 °C step increases the pressure by another 1000 bars. Temperatures in this range would not have been sufficient to completely obliterate remnants of microfossils and indigenous organic substances. Pressure probaby would not have destroyed microfossils either, since the microstructures are surrounded by quartz, which protected them against deformation and destruction. This mode of preservation would effectively negate arguments based on the unlikely chances of survival of such "delicate structures" through the several metamorphic events. Once a microstructure becomes enclosed within a single crystal of quartz, deformation of the surrounding rocks would be of no consequence. Graphitisation of the organic matter is incomplete under the given conditions because the gaseous products cannot escape.

A study of the Isua organic matter is still in progress (McKirdy & Hahn, 1982) [5]. Volatile organics have been found to occur chiefly in quartzites and in silicate-rich BIF. Alkane distributions show great variability, even over distances of a

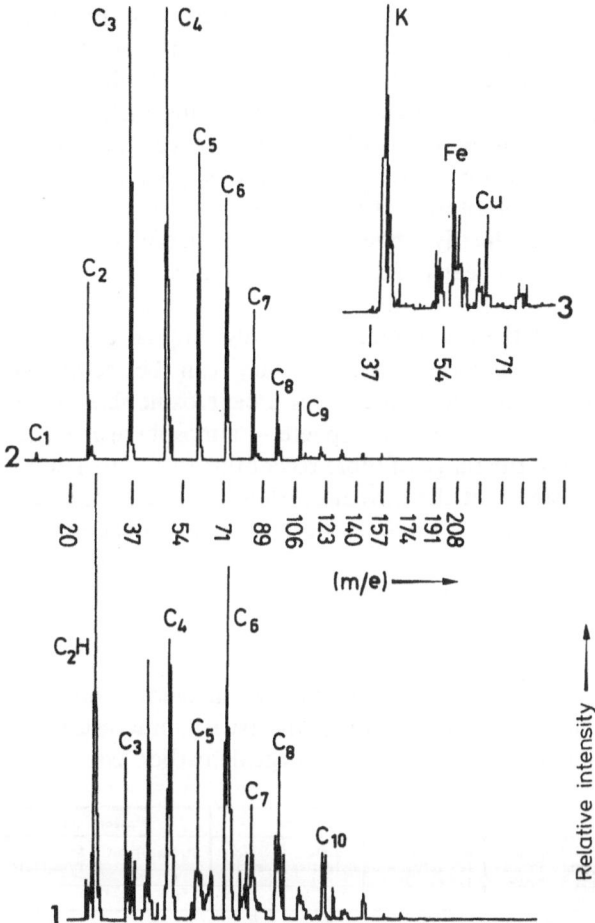

Fig. 39. Laser mass spectra of a microstructure corresponding to Fig. 38 analyzed in thin Section in situ. 1. Negative ions, 3. Positive ions 2. Spectrum of artificial graphite (Pflug, 1984b) [124]

few meters. Some samples appear to be contaminated by organic matter of relatively recent origin, others by older epigenetic materials brought into contact with the rocks by glacial action or by contamination with petroleum products. However, several rock samples have yielded an alkane pattern interpreted to contain syngenetic microbial organic matter at a rather late stage of catagenesis, an observation which is consistent with the amphibolite — grade metamorphism of the Isua terrain.

Carbon isotope ratios from Isua carbonate horizons showed a mean $\delta^{13}C = -2.5\%_{00}$ for carbonate carbon, and $\delta^{13}C$ values for graphitic carbon widely scattered, from $-7\%_{00}$ to $-22\%_{00}$. Schidlowski (1984a, b) [2,97] interpreted these carbon isotopic ratios as having been altered during metamorphism from initially $\delta^{13}C_{carb.} = \pm 0\%$ and $\delta^{13}C_{graph-} = -25\%_{00}$, thereby indicating that life processes probably were active 3.8 Ga ago. Since the suggested isotope shifts are consistent with an isotopic reequilibration between organic carbon and carbonate in response to amphibolite-grade metamorphism, it is reasonable to assume that the isotopic signature of autotrophic carbon fixation originally extended to 3.8 Ga ago. Consistent with this argumentation is a statement of Sidorenko (1984) [55] implying that even highly aluminiferous schists and gneisses of the granulite metamorphic facies in Kola Peninsula (a petrochemical reconstruction has proved their "para-nature") contain carbonaceous matter up to 1.5% with isotope-proved biogenic nature. According to Schidlowski's quantitative estimates, the sedimentary organic carbon reservoir at Isua times was already about 50–60% as large as it is at present. The main precursors of this early Precambrian organic matter were probably primitive autotrophs.

In summary, it can be stated that there is nothing markedly unusual about the Isua supracrustal facies and their organic contents. They can be compared closely with younger Precambrian, and in some cases Phanerozoic lithologies and successions. Depositional mechanisms and hydrosphere — atmosphere chemical equilibria appear to have been within the range of more recent times (Nutman et al., 1984) [119]. It seems from all indication that the advent of photoautrophy preceeded Isua times. It was probably the most crucial single event to make an impact on the evolution of the terrestrial atmosphere.

X Final Remarks

There exists a continuous record of organic microstructures in ancient minerals and sediments from the ~3.8 Ga old metasediments of the Isua Supracrustal Belt (Fig. 40). It has been pointed out that variations in the organic carbon content of

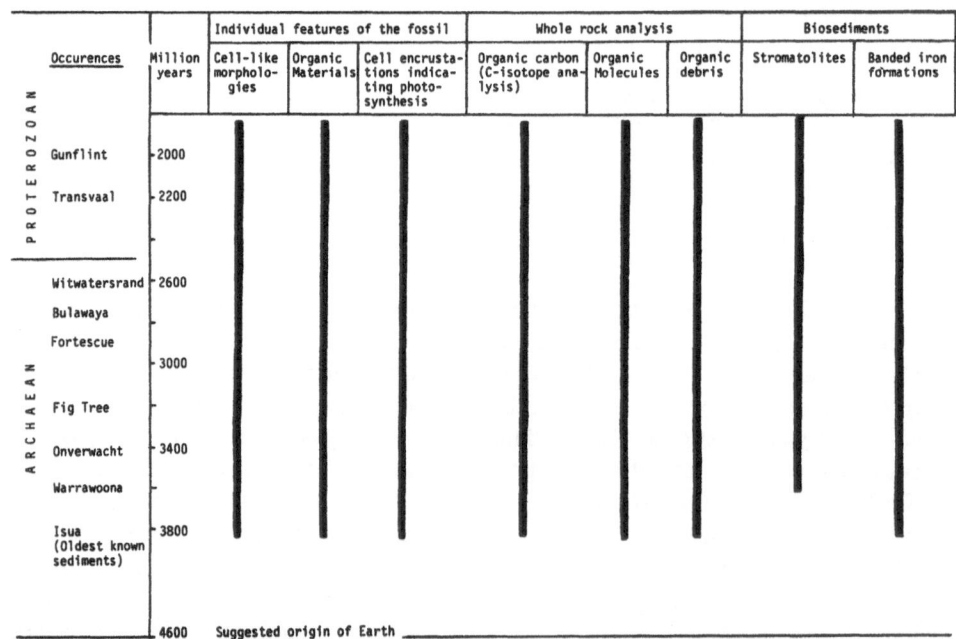

Fig. 40. Indication of life in the Precambrian prior to 2000 MY. Left column: Stratigraphic position of the cited occurrences (Pflug, 1982) [16]

Archaean rocks lie well within the realm of geologically younger sediments, often approaching means between 0.5 and 0.6% (Schidlowski, 1984a) [97]. This range which has been proposed as the best current approximation for the average organic carbon content of the sedimentary shell as a whole. Although magmatic rocks may contain organic carbon of non-biological origin (Freund et al., 1982) [124] no syntheses other than biological have as yet been shown to proceed in terrestrial near-surface environments. Thus, the presence of kerogenous particles and their metamorphosed derivatives in ancient minerals would constitute, per se, evidence of an early advent of biological carbon fixation.

When did life start? This question was discussed in detail at a recent symposium (Moorbath, 1983) [125]. Nobody seriously considered a date younger than 4000 MY to be realistic. The consensus was based on a cautious interpretation of the evidence from Isua. Others prefered to base their opinion on the general evidence for the cessation of extraterrestrial bombardment at around 4000–3900 MY ago.

So, in the absence of any evidence to the contrary, it must be concluded that life on Earth is considerably older than the oldest known rocks. Consequently, it may not be possible on Earth to find indication for any of the prebiological processes that chemists have so often attempted to simulate in the laboratory. Evidence for such processes may have to be looked for in other places, such as interstellar space, meteorites, and the surfaces of other planets (Ponnamperuma, 1983) [126].

XI Acknowledgements

I am indebted to P. Haug, Mainz, B. Heinz, Frankfurt, F. Hillenkamp, Frankfurt for reading the manuscript and expertal advice.

XII References

1. Schidlowski, M., Appel, P. W. U., Eichmann, R., Junge, C. F.: Geochim. Cosmochim. Acta 43, 189 (1979)
2. Schidlowski, M.: Proc. 27th Internat. Geol. Congr. Moscow 19, 229 1984b
3. Eglington, G.: Organic molecules as chemical fossils — the molecular fossil record, in: Cosmochemistry and the origin of life, in: Cosmochemistry and the origin of life (ed. Ponnamperuma, C.) p. 323, Dordrecht, Boston, London, D. Reidel, P. C. 1983
4. Leventhal, J.: Proc. Nat. Acad. Sci. USA 72, 4706 (1975)
5. McKirdy, D. M., Hahn, J. H.: The composition of kerogen and hydrocarbons in Precambrian rocks, in: Mineral deposits and the evolution of the biosphere (eds. Holland H. D., Schidlowski, M.) p. 123, Berlin Heidelberg, New York, Tokio, Springer 1982
6. Philp, R. P., van de Ment, D.: Precambrian Research 20, 3 (1983)
7. Schopf, J. W., Walter, M. R.: Archean microfossils: New evidence of ancient microbes, in: Earth's earliest biosphere: Its origin and evolution (ed. Schopf, J. W.) p. 214, Princeton N.J., Princeton University Press 1983
8. Klemm, D. D.: Mineral. Deposita 14, 381 (1979)
9. Huestis, L.: J. Chem. Ed. 53, 270 (1976)
10. Schmitt, J. G., Boyd, D. W.: J. Sediment Petrol. 51, 1297 (1981)
11. Sigleo, A. C.: Geochim. Cosmochim. Acta 42, 1397 (1978)

12. Leo, R. F., Barghoorn, E. S.: Harvard Univ. Bot. Mus. Leaflets *25*, 1 (1976)

13. Merill, R. W., Spencer, R. C.: Ind. Eng. Chem. *42*, 744 (1950)

14. Allen, E. T.: Am. J. Sci. *28*, 373 (1934)

15. Francis, S., Margulis, L., Barghoorn, E. S.: Precambrian Research *6*, 65 (1978)

16. Pflug, H. D.: Zbl. Bakt. Hyg., I, Abt. Orig. *C 3*, 53 (1982)

17. Pflug, H. D.: Ultrafine structure of the organic matter in meteorites, in: Fundamental studies and the future of science (ed. C. Wickramasinghë, C.) p. 24, Cardiff, University Press 1984a

18. Pflug, H. D.: Unpublished data (1985)

19. Obérlin, A. et al.: Electron microscopic study of kerogen microtexture, in: Kerogen, (ed. Durand, B.) p. 191, Paris, Technip 1980

20. Speight, J. G.: Applied Spectroscopy Rev. *5*, 211 (1971)

21. Robin, P. L. et al.: Characterisation des kérogèns par spectroscopie infrarouge, in: Advances in organic geochemistry "1975", (eds. Campos, R., Goni, J.) p. 693, Madrid, Enadisma 1977

22. Rouxhet, P. G. et al.: Characterization of kerogens and their evolution by infrared spectroscopy, in: Kerogen (ed. Durand, B.) p. 163, Paris, Technip 1980

23. Friedel, R. A.: Proc. 4th Carbon Conf. London, 321 (1960)

24. Tissot, B. P., Welte, D. H.: Petroleum formation and occurence, Berlin, Heidelberg, New York, Tokyo, Springer 1978

25. Friedel, R. A., Carlson, G. L.: Fuel *51*, 194 (1972)

26. Robin, P. L.: Charactérisation de kérogènes et de leur évolution par spectroscopie infra-rouge. Thesis, Univ. Louvain 1975

27. Filip, Z.: Infrared spectroscopy of two soils and their components, in: Environmental biogeochemistry and geomicrobiology (ed. Krumbein, W. E.) p. 747, Ann Arbor Mich., Ann Arbor Sci. 1978

28. Hunt, J. M. et al.: Analytical Chemistry *22*, 1478 (1950)

29. Chester, R., Elderfield, H.: Geochim. Cosmochim. Acta *32*, 1128 (1968)

30. Liese, H. C., Selected terrestrial minerals and their infrared absorption spectral data 4000 — 300 cm^{-1}, in: Infrared and Raman spectroscopy of lunar and terrestrial materials (ed. Karr, C.) p. 197, New York, S. Francisco, London, Academic Press 1975

31. Griffith, W. P.: Raman spectroscopy of terrestrial minerals, in: Infrared and Raman spectroscopy of lunar and terrestrial materials (ed. Karr, C.) p. 299, New York, S. Francisco, London, Academic Press 1975

32. van der Marel, H. W., Beutelspacher, H.: Atlas of infrared spectroscopy of clay minerals and their admixtures. Amsterdam, Oxford, New York, Elsevier 1976

33. Southworth, D.: Grana Palynologica *9*, 5 (1969)

34. McCartney, J. T., Ergun, S.: US Dept. Interior Bureau of Mines Bull. *641*, 28 (1967)

35. Nagy, L. A.: J. Paleontology *52*, 141 (1978)

36. Friedel, R. A., Retcofsky, H. L.: in: Spectrometry of Fuels (ed. Friedel, R. A.) p. 46, New York—London, Plenum Press 1970

37. Retcofsky, H. L., Friedel, R. A.: Spectra of coals and coal extracts. — Ultraviolet-visible spectra of carbon disulfide extracts, in: Spectrometry of Fuels (ed. Friedel, R. A.) p. 37, New York—London, Plenum Press 1970

38. Bell et al.: Absorption spectroscopy of ionic and molecular units in crystals and glasses, in: Infrared and Raman spectroscopy of lunar and terrestrial materials (ed. Karr, C.) p. 1, New York, S. Francisco, London, Academic Press 1975

39. Hillenkamp, F., Feigl, P., Schüler, B.: Laser micromass analysis of bulk surfaces, in Microbeam Analysis (ed. Heinrich, K. F. J.) p, 359, San Francisco, San Francisco Press 1982

40. Simons, D. S.: Laser microprobe mass spectrometry, in: Springer Series in Chemical Physics *36* (eds. Benninghoven, A.), et al., p. 158, Berlin, Heidelberg, New York, Tokyo, Springer 1984

41. Kaufmann, R.: LAMMA, Reference and Abstract Index, Univ. Düsseldorf, Dept. Clinic. Physiol. 1984

42. Gardella et al.: Spectroscopy Letters *13*, 347 (1980)

43. Fürstenau, N., Hillenkamp, F., Nitsche, R.: Int. J. Mass Spectr. and Ion Phys. *31*, 85 (1979)

44. Mel'nik, Y. P.: Precambrian Banded Iron-Formations, Amsterdam, Oxford, New York, Elsevier Sci. P.C. 1982
45. Muir, M. D.: Microenvironments of some modern and fossil iron- and manganese-oxidizing bacteria, in: Environmental biogeochemistry and geomicrobiology 3 (ed. Krumbein, W. E.) p. 937, Ann Arbor Mich; Ann Arbor Publishers Sci. 1978
46. Cowen, J. P.: Fe and Mn depositing bacteria in marine suspended macroparticulates, in: Biomineralization and biological metal accumulation (eds. Westbroek, P., de Jong, E. W.) p. 489, Dordrecht, Boston, London, D. Reidel P.C. 1983
47. Golubic, S.: Stromatolites, fossil and recent: a case history, in: Biomineralization and biological metal accumulation (eds. Westbroek, P., de Jong, E. W.) p. 313, Dordrecht, Reidel Publ. Co., 1983
48. Awramik, S. M.: Ancient stromatolites and microbial mats, in: Microbial mats: Stromatolites (eds. Cohen, Y., et al.) p. 1, New York, Alan R. Liss Inc. 1984
49. Monty, C. L. V.: Terra Cognita 4, 423 (1984)
50. Krumbein, W. E.: Algal mats and their lithification, in: Environmental biogeochemistry and geomicrobiology, 1 (ed. Krumbein, W. E.) p. 209, Ann Arbor Mich; Ann Arbor Publishers Sci. 1978
51. Lyons, W. B. et al.: Geology 12, 623 (1984)
52. Boon, J. J.: Tracing the origin of chemical fossils in: microbial mats: Biogeochemical investigations of Solar Lake cyanobacterial mats using analytical pyrolysis methods, in: Microbial mats: Stromatolites (eds. Cohen, Y., et al.) p. 318, New York, Alan R. Liss Inc. 1984
53. Durand, B.: Kerogen, Paris, Editions Technip 1980
54. Hayes, J. M., Kaplan, I. R., Wedeking, K. W.: Precambrian organic geochemistry, preservation of the record, in: Earth's earliest biosphere: Its origin and evolution (ed. Schopf, J. W.) p. 93, Princeton N.J., Princeton University Press 1983
55. Sidorenko, S. A.: Proc. 27th Int. Geol. Congr. Moscow 5, 199 (1984)
56. Raynaud, J. F., Robert, P.: Bull. Centre Rech. Pau-SNPA 10, 109 (1976)
57. Peters, K. E., Ishwatari, R., Kaplan, I. R.: Amer. Assoc. Petrol. Geol. Bull. 61, 504 (1977)
58. Staplin, F.: Determination of thermal alteration index from color of exinite (pollen, spores), in: How to assess maturation and paleotemperatures, (eds. Staplin, F. L., et al.) SEPM Short Course 7, p. 91, Tulsa OK, SEPM 1982
59. Heroux, Y., Chagnon, A., Bertrand, R.: Amer. Assoc. Petrol. Geol. Bull. 63, 2128 (1979)
60. Pflug, H. D.: Geologica et Palaeontologica 13, 1 (1979)
61. Brown, J. K.: J. Chem. Soc. "1955", 744 (1955)
62. Fujii, S. et al.: Fuel 49, 68 (1970)
63. Cloud, P. E., Morrison, K.: Precambrian Research 9, 81 (1979)
64. Karavaiko, G. I.: Microflora of land microenvironments and its role in the turnover of substances, in: Environmental biogeochemistry and geomicrobiology 2 (ed. Krumbein, W.-E.) p. 397, Ann Arbor Mich; Ann Arbor Sci. Publishers Inc. 1978
65. Nagy, B., Engel, M. H., Zumberge, J. E., Ogino, H., Chang, S. Y.: Nature 289, 53 (1981)
66. Buick, R.: Precambrian Research 24, 157 (1984)
67. Smith et al.: Geochim. Cosmochim. Acta 34, 659 (1970)
68. Oehler, J. H.: Precambrian Research 4, 221 (1977)
69. Teichmüller, M., Durand, B.: Int. J. Coal Geol. 2, 197 (1983)
70. Van Gijzel, P.: Characterization and identification of kerogen and determination of thermal maturation by means of qualitative and quantitative microscopical techniques, in: How to assess maturation and paleotemperatures (eds. Staplin, F. L. et al.) SEPM Short Course 7, p. 91, Tulsa OK, SEPM 1982
71. Teichmüller, M., Wolf, M.: J. Microscopy 109, 49 (1977)
72. Kirschvink, J. L., Chang, S. R.: Geology 12, 559 (1984)
73. Lowenstam, H. A.: Science 211, 1126 (1981)
74. Lowenstam, H. A.: Proc. 27th Internat. Geol. Congr. Moscow 2, 79 (1984)
75. Jones, G. E. et al.: Trace element composition of five Cyanobacteria, in: Environmental biogeochemistry and geomicrobiology 3 (ed. Krumbein, W. E.) p. 967, Ann Arbor Mich, Ann Arbor Science 1978

76. Ochiai, E.: Inorganic chemistry of earliest sediments, in: Cosmochemistry and the origin of life (ed. Ponnamperuma, C.) p. 235, Dordrecht, Boston, London, Reidel P.C. 1983
77. Udel'nova, T. M., Gnilovskaya, M. B., Boychenko, Y. A.: Dokl. Acad. Sci. USSR Earth Sci. Sct. (Washington DC) (260/1–6), 187 (1983)
78. Kelly, W. C., Nishioka, G. K.: Geology 13, 334 (1985)
79. Hoering, T. C.: Carnegie Inst. Washington Yearbook 75, 806 (1976)
80. Westbroek, P.: Biological metal accumulation and biomineralization in a geological perspective, in: Biomineralization and biological metal accumulation (eds. Westbroek, P., de Jong, E. W.) p. 1, Dordrecht, Boston, London, D. Reidel P.C. 1983
81. Lein, A. Y.: Formation of carbonate and sulfide minerals during diagenesis of reduced sediments, in: Environmental biogeochemistry and geomicrobiology 1 (ed. Krumbein, W. E.) p. 339, Ann Arbor Mich. Ann Arbor Publishers Sci. 1978
82. Friedman, G. M.: Importance of microorganisms in sedimentation, in: Environmental biogeochemistry and geomicrobiology 1 (ed. Krumbein, W. E.) p. 323, Ann Arbor Mich, Ann Arbor Publishers Sci. 1978
83. Berner, R. A.: Science 159, 195 (1968)
84. Kazmierczak, J.: Precambrian Research 9, 1 (1979)
85. Porter, K. G., Robbins, E. L.: Science 212, 931 (1981)
86. LaBerge, G. L., Robbins, E. I., Schmidt, R. G.: Abstracts 27th Internat. Geol. Congr. Moscow 1, 241 (1984)
87. Schidlowski, M., Hayes, J. M., Kaplan, I. R.: Isotopic inferences of ancient biochemistries: Carbon, sulfur, hydrogen, and nitrogen, in: Earth's earliest biosphere: Its origin and evolution (ed. Schopf, J. W.) p. 149, Princeton, N. J., Princeton University Press 1983
88. Casagrande, D., Siefert, K.: Science 195, 675 (1977)
89. Berner, R. A.: Geochim. Cosmochim. Acta 48, 605 (1984)
90. Kalliokoski, J.: Economic Geology 61, 872 (1966)
91. Stürmer, W.: Interdisciplinary Sci. Rev. 9, 1 (1984)
92. Tufar, W.: Mitt. österr. geol. Ges. 77, 185 (1984)
93. Ramdohr, P.: Geol. Jahrb. 67, 367 (1953)
94. Harder, E. B.: US Geol. Surv. Prof. Pap. 43, 1 (1919)
95. Schopf, J. M., Ehlers, E. G., Stiles, D. V., Birle, J. D.: Proc. Amer. Phil. Soc. 109, 288 (1965)
96. Love, L. G., Zimmerman, D. O.: Econ. Geol. 56, 873 (1961)
97. Schidlowski, M.: J. Geol. Soc. London 141, 243 (1984a)
98. Barghoorn, E. S., Tyler, S. A.: Science 147, 563 (1965)
99. Lougheed, M. S.: Geol. Soc. Amer. Bull. 94, 325 (1983)
100. Krumbein, W. E., Barghoorn, E. S., Knoll, A. H.: Terra Cognita 4 (1984)
101. Robin, P. L., Rouxhet, P. G.: Revue Inst. Franc. Petrol. 31, 955 (1976)
102. Yen, T. F., Erdmann, J. G.: Preprint General Pap. Division Petrol. Chem., — Amer. Chem. Chem. Soc. 7, 5 (1962)
103. King, L. H. et al.: Dept. of Mines, Technical Surv., Fuels Mining Practice Div. R 114, 1 (1963)
104. LaBerge, G. L.: Geol. Soc. Amer. Bull. 78, 331 (1967)
105. Gross, G. A.: Sediment. Geol. 7, 241 (1972)
106. Nagy, L. A., Zumberge, J. E.: Proc. Nat. Acad. Sci. 73, 2973 (1976)
107. Sklarew, D., Nagy, B.: Proc. Natl. Acad. Sci. USA 76, 10 (1979)
108. Pflug, H. D.: Die Spur des Lebens, Berlin—Heidelberg—New York—Tokyo, Springer 1984c
109. Nagy, B.: Naturwissenschaften 69, 301 (1982)
110. Hofmann, H. J.: Geol. Surv. Canada, Pap. 84, 285 (1984)
111. Nagy, B.: Naturwissenschaften 63, 499 (1976)
112. Hoefs, J., Schidlowski, M.: Science 155, 1096 (1967)
113. Hallbauer, D. K.: Minerals Sci. Engng. 7, 111 (1975)
114. Dexter-Dyer Grosovsky, B.: Microbial role in Witwatersrand gold deposition, in: Biomineralization and biological metal accumulation, (eds. Westbroek, P. de Jong, E. W.) p. 495, Dordrecht, D. Reidel, Publ. Co. 1983
115. Rouzaud, J. N., Oberlin, A., Trichet, J.: Interaction of uranium and organic matter in uraniferous sediments, in: Advances of organic Geochemistry (eds. Douglas, A. G. Maxwell, J. R.) p. 505, Oxford, Pergamon Press 1982
116. Pflug, H. D.: Rev. Palaeobotan. Palynol. 5, 9 (1967)

117. Pflug, H. D., Jaeschke-Boyer, H., Sattler, E. L.: Microscopia Acta *82*, 255 (1979)
118. Appel, P. W. U.: Precambrian Research *11*, 73 (1980)
119. Nutman, A. P. et al.: ibid. *25*, 365 (1984)
120. Nagy, B., Zumberge, J. E., Nagy, L. A.: Proc. Nat. Acad. Sci. USA *72*, 1206 (1975)
121. Bridgwater, D. et al.: Nature *289*, 51 (1981)
122. Roedder, E.: ibid. *293*, 459 (1981)
123. Pflug, H. D.: Naturwissenschaften *71*, 63 (1984b)
124. Freund, F. et al.: Geol. Rundsch. *71*, 1 (1982)
125. Moorbath, S.: The dating of the earliest sediments on Earth, in: Cosmochemistry and the origin of life (ed. Ponnamperuma, C.) p. 213, Dordrecht, Boston, London, D. Reidel P. C. 1983
126. Ponnamperuma, C.: Cosmochemistry and the origin of life, in: Cosmochemistry and the origin of life (ed. Ponnamperuma, C.) p. 1, Dordrecht, Boston, London, D. Reidel P. C. 1983

Metabolism of Proteinoid Microspheres

Tadayoshi Nakashima

Institute for Molecular and Cellular Evolution University of Miami, Coral Gables, Florida, USA

Table of Contents

The literature of metabolism in proteinoids and proteinoid microspheres is reviewed and criticized from a biochemical and experimental point of view. Closely related literature is also reviewed in order to understand the function of proteinoids and proteinoid microspheres. Proteinoids or proteinoid microspheres have many activities. Esterolysis, decarboxylation, amination, deamination, and oxido-reduction are catabolic enzyme activities. The formation of ATP, peptides or oligonucleotides is synthetic enzyme activities. Additional activities are hormonal and inhibitory. Selective formation of peptides is an activity of nucleoproteinoid microspheres; these are a model for ribosomes. Mechanisms of peptide and oligonucleotide syntheses from amino acids and nucleotide triphosphate by proteinoid microspheres are tentatively proposed as an integrative consequence of reviewing the literature.

1 Introduction

Proteinoids, as a model of primitive abiotic proteins [1], are formed by polymerization from protobiologically plausible micromolecules (amino acids) under presumed protobiological conditions. Proteinoids have enzyme-like activities and metabolic qualities. Proteinoid microspheres are the most suitable model for protocells since they do not consist of macromolecules extracted from contemporary organisms.

A number of studies have been performed in the context of a theory that proteins and polynucleotides were formed in a suspension of proteinoid microspheres; and the microspheres could then have evolved to contemporary cells. The experimental results and evolutionary considerations have been summarized in the textbook of Fox and Dose published in 1977 [2]. This review therefore deals with studies since 1977, although some description of literature before 1977 is reviewed as occasion demands. Since the evolutionary consideration of proteinoids and proteinoid microspheres has been discussed in much literature and many books, (e.g.: 2, 3), the attention in this paper is focussed on the description of the biochemical and experimental parts of the literature. Inasmuch as protobiological activities of proteinoids in solution are carried into microspheres [2], experiments with proteinoids in solution are not excluded.

Proteinoids, which are made from a mixture of amino acids by heating at temperatures in the range of 120–200 °C, contain pigments and other by-products besides the main product, peptides. It is a formidable task to analyze proteinoids thoroughly, because of by-products that bind or otherwise interact with mixed polymers. The determination of amino acid composition and the terminal amino acids of proteinoids has been done without difficulty. Primary structure of proteinoids is not known, except for the amino acid sequence of two tripeptides [4] and a number of terminal residues. In this context, the active sites of proteinoid are suggested only from their amino acid compositions. The dynamic character of proteinoids makes their handling difficult and impairs the reproducibility of results. Consequently, there are a number of unpublished experimental results. In the studies of the metabolism by proteinoid microspheres, every experiment is done with prebiologically plausible materials and environments (for example in aqueous solution, not in organic solvent to synthesize plausible products).

In this context, the literature concerning closely relevant experiments and the understanding of mechanisms of metabolism in proteinoid microspheres are reviewed.

2 Proteinoid Microspheres

The proteinoid microspheres (Fig. 1), as simulated protocells, form from virtually all of the known wide variety of thermal copolyamino acids [2]. Microspheres are formed if the aqueous or aqueous salt solution of proteinoid is heated and the clear decanted solution is allowed to cool. This self-assembly may also be effected by chilling solutions saturated at room temperature. Sonication at room temperature can be used.

The shape of the microspheres is usually spheroidal; each population tends to be uniform. The sizes tend to fall in the range 0.5–7.0 µm in diameter and the operational factors controlling size of the microspheres appear to be many: kind of proteinoid,

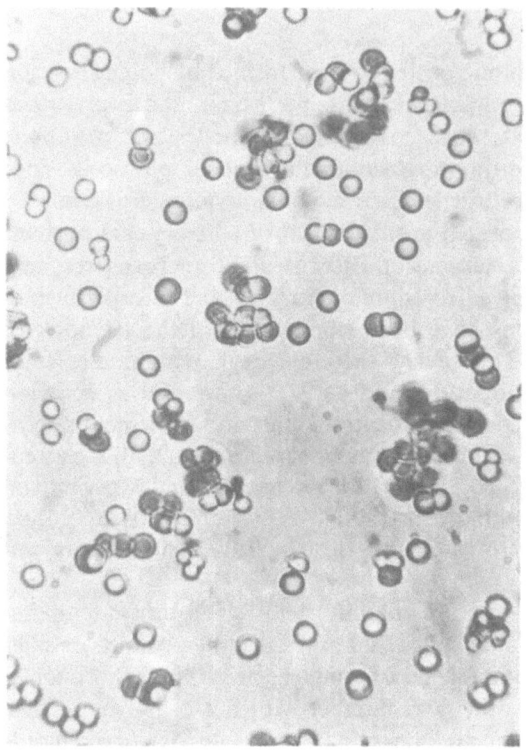

Fig. 1. Proteinoid microspheres: Fox, Dose [2]

added substances, temperature of heated solution, etc. [2]. The microspheres composed of both acidic proteinoid (e.g. glutamic acid-rich) and basic proteinoid (e.g. lysine-rich) are stable at high temperature. In this kind of assembly, interaction between polyanion and polycation must be involved [3]. Incorporation of catalytic functions of the proteinoids into microspheres assembled therefrom should be anticipated [3].

When the pH of a suspension of microspheres of acidic proteinoid is raised by 1–2 units, diffusion of material from the interior to the exterior, fission into two particles, and the appearance of a double layer in the boundary are observed [2]. Proteinoid microspheres shrink or swell on transfer to hypertonic or to hypotonic solutions respectively. Some experiments show that polysaccharides are retained under conditions in which monosaccharides diffuse out [2]. Some proteinoid microspheres possess the intrinsic capacity to grow by accretion, to proliferate through budding, and to form junctions [2]. The morphology and other characteristics of proteinoid microspheres are altered by the inclusion of other materials such as polynucleotides, lipids or salts.

3 Catabolic and Synthetic Enzyme Activities of Proteinoids and Proteinoid Microspheres

Esterolysis, decarboxylation, amination, deamination and oxidoreduction are catabolic enzyme activities of proteinoids and proteinoid microspheres. The formation of ATP, peptides or oligonucleotides are known synthetic enzyme activities of proteinoid microspheres. The literature of esterolysis, decarboxylation, amination deamination and oxidoreduction catalyzed by proteinoids has been reviewed [5, 6].

3.1 Esterolysis

3.1.1 Hydrolysis of *p*-nitrophenyl Acetate

Acidic proteinoids accelerate the hydrolysis of the unnatural substrate, *p*-nitrophenyl acetate [7, 8]. *P*-Nitrophenyl acetate has been used as a substrate for both natural esterases and esterase models. The imidazole ring of histidine is involved in the active site of a variety of enzymes, including hydrolytic enzymes. Histidine residues of proteinoid play a key role in the hydrolysis, the contribution to activity of residues of lysine and arginine is minor, and no activity is observed for proteinoid containing no basic amino acid [7].

The proteinoids are inactivated by heating in buffer solution or by treatment with alkali at room temperature, and it is proved that the hydrolysis of cyclic imide bonds, in which aspartic acid residues are initially bound, accompanies the inactivation by heat [8].

A typical reaction mixture for hydrolysis contained 3.6 mg of *p*-nitrophenyl acetate (2 mM), 2% in acetone, and 4.0 mg of proteinoid in 10 ml of phosphate buffer, pH 6.2, 31–32 °C. The liberated *p*-nitrophenol was assayed with a spectrophotometer (400 nm) [7].

Several kinds of evidence indicate that the reactions are catalytic rather than stoichiometric. When the reaction is followed to completion, linear first order plots are obtained for at least 90% of the reaction [7]. At the ratio of substrate to polymer employed, about 1:1 by weight, nonlinear first order plots would be predicted for a stoichiometric reaction. When a second aliquot of substrate is added after completion of the reaction, the first order rate constant noted with the second aliquot is essentially identical to that of the original [7]. The liberation of acetate and *p*-nitrophenol in equimolar proportions is also consistent with an inference of catalysis [7].

3.1.2 Hydrolysis of Phosphate Ester

Alkali-treated proteinoids containing the 18 common amino acids promote the hydrolysis of the ester bond of *p*-nitrophenylphosphate [9]. In general, the higher the proportion of neutral and basic amino acids proteinoid has, the higher the activity is [9].

The reaction is carried out in 3.3 ml of 0.03 M Tris buffer, pH 7.6, containing 3.3 nmoles of $ZnCl_2$, 33 nmoles $MgCl_2$, 3.3 μmoles of substrate, and 0.1 to 5.0 mg of proteinoid at 30 °C. The released *p*-nitrophenol is measured spectrophotometrically at 410 nm [9].

The rate of hydrolysis of *p*-nitrophenylphosphate by a fraction of neutral proteinoid separated by gel filtration is 3 μmoles/day/mg at 30 °C, while cow milk protein has an activity 15–31 μmoles/day/mg at 37 °C [9].

The characteristics of the proteinoid activity resemble those of natural enzymes in several respects. The activity of the proteinoid is inhibited by typical phosphatase inhibitors such as arsenate or phosphate, the effects of divalent metal ions are similar, and the optimum pH value of 6.7 is between those of alkaline and acid phosphatases [9]. The K_m value (1.4×10 M) of the proteinoid is about 10 times greater than those of natural phosphatases, e.g. alkaline phosphatase from *E. coli* has a K_m value 1.2×10 M [9].

3-O-Methyl fluorescein phosphate is also hydrolysed by the proteinoids [9].

An active proteinoid fraction which was separated by gel filtration is almost totally inactivated by heating a solution at 80 °C for 5 minutes [9]. The fact that similar heat treatment during the processing of the proteinoid did not result in loss of activity suggests that the gel filtration step may have removed an unknown stabilizing factor [9].

3.2 ATP Protometabolism

3.2.1 Hydrolysis of ATP

Proteinoid microspheres containing zinc hydrolyze the natural substrate, adenosine triphosphate (ATP) as well as the unnatural substrates, *p*-nitrophenylacetate or *p*-nitrophenyl phosphate. The significance resides in the fact that the energy for most biosyntheses is provided by the hydrolysis of ATP. Zinc, magnesium and other metal salts are known to catalyze the hydrolysis of ATP [10]. Proteinoid microspheres containing zinc as a cofactor have an activity for hydrolysis of ATP [11, 12].

After the cooling of a hot solution of acidic proteinoid and zinc hydroxide, zinc-containing microspheres are deposited. The Zn-proteinoid microspheres washed with water retain the activity, but the wash liquids show successively less activity [11]. Attempts to introduce Zn into proteinoids in this manner gave highly erratic results until freshly prepared zinc hydroxide was used [11]. There have not been further experimental data published; these experiments were mainly to attempt introducing zinc into proteinoid microspheres.

3.2.2 ATP Synthesis

Conversion of ADP to ATP has been accomplished in aqueous suspensions of thermal Poly(Asp, Glu, Cys, Leu, Tyr) microspheres. Tyrosine was included in the attempt

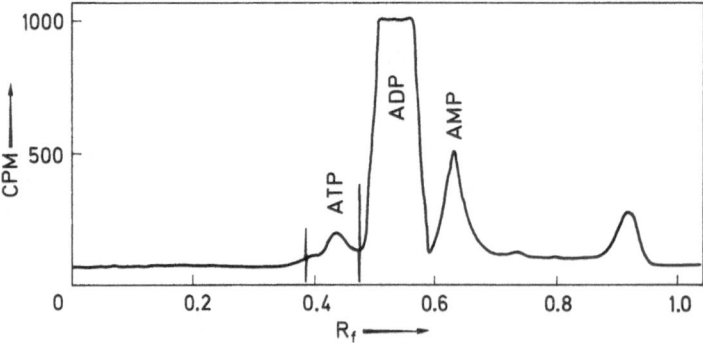

Fig. 2. Chromatogram showing ATP produced in a suspension of proteinoid microspheres in an aqueous solution of ADP and inorganic phosphate: Fox, Adachi, Stillwell [13]

to provide a nascent structure in the proteinoid susceptible to conversion to the corresponding quinone. Tyrosine is known to undergo several stages of oxidation toward quinone [13]. After illumination by white light for 14 hours, a reaction mixture of the microspheres described (15 mg), $FeCl_2$ 4 H_2O (20 mg), KH_2PO_4 (40 mg) and ADP (8–[14]C) sodium salt (100 mg) is chromatographed. The radiochromatogram reveals a peak in the ATP region (Fig. 2) [13]. The yield of ATP is 0.15–0.25%. The quantum yield must also be very small; however, the yield corrected for the small volume of microspheres would be considerably larger [13]. The iron salt is found to promote the production of the ATP, the optimal pH is between 2 and 3. The pH values of 5, 6, and 7 were also tested, but no yield of ATP was observed in that range [13].

The conversion occurs also in an aqueous suspension of proteinoid microspheres containing dopa in the polymer instead of tyrosine. Presumably, dopaquinone is formed from tyrosine or dopa within the proteinoid during illumination with the aid of ferrous chloride [13]. These experiments were performed after studies of the conversion of ADP to ATP catalyzed by the quinone, chloranil, and ferrous ion in dimethylformamide solution with white light. The yield of ATP under these condition is 20% [13, 14].

In related studies of metal ions, Krasnovski learned that the oxides of titanium, zinc, or tungsten possess high photosensitizing activity in redox reactions comparable to the activity of porphyrins and chlorophylls [15].

The experiments to utilize proteinoid catalysts for efficient direct photophosphorylation of ADP to ATP in water solution are still in a preliminary stage of development. For example, the following reports might help further investigation. When adenyl-5'-yl phosphoramidate is incubated with orthophosphate in aqueous solution at pH 3 for 48 hrs, 40% of adenyl-5'-yl phosphoramidate is converted to ATP [16]. The diverse set of enzymes that build up and break down ATP may work according to the same mechanism. All of these enzymes can exist in two conformational forms, which catalyze ATP synthesis and hydrolysis [17].

3.3 Decarboxylation

3.3.1 Oxaloacetic Acid

Proteinoids catalyze the decarboxylation of oxaloacetic acid to pyruvic acid, lysine-rich proteinoid being the most effective of these tested. Some are about 15 times more active than the equivalent amount of free lysine. Acidic proteinoids exhibit very little activity [18].

A typical reaction mixture contains 1.0–4.0 mg of proteinoid and approximately 10 µmoles (1.32 mg) of substrate, in 2.1 ml of 0.2 M acetate buffer, pH 5.0 (the optimum pH). Liberated CO_2 is trapped in KOH solution. The reactions are usually followed to about 80–90% completion, which requires 1–2 hours or less. This reaction, under conditions yielding pseudo first-order kinetics, is catalytic [18].

The activity of the proteinoids is not due to metal ion contaminants. Analyses of samples of lysine-rich proteinoid indicated very low levels of metals (ca. 0.1 µmole/mg), possibly attributable to handling during analysis [18]. Amino groups of proteinoids are involved in the catalytic process, as indicated by the fact that acetylated proteinoids are nearly devoid of activity.

Proteinoids were tested after being stored in the dry state for 5 to 10 years. Acidic proteinoids effective in catalyzing the hydrolysis of p-nitrophenyl acetate showed the same levels of activity as observed 10 years earlier [19]. Lysine-rich proteinoids which catalyzed the decarboxylation of oxaloacetic acid were found to be insoluble in assay medium after 5 years of storage in the dry state. Their activity, however, had increased by 32 to 145%, and the activity of the lysine-rich proteinoids was largely associated with the insoluble portion [19].

3.3.2 Decarboxylation of Pyruvic Acid

Proteinoids accelerate the conversion of pyruvic acid to acetic acid and carbon dioxide. In a typical experiment, a mixture of 20 mg of proteinoid and 0.4 mg of radioactive pyruvic acid in 20 ml of 0.2 N Tris buffer, pH 8.3, is incubated at 37.5 °C for 1, 2 or 3 days. The evolved $^{14}CO_2$ is absorbed in NaOH solution, and acetic acid is identified by paper chromatography and by preparing a derivative of acetic acid, namely p-bromophenacyl acetate [20].

In general, acidic proteinoids are more active than lysine-rich proteinoids for this reaction. Thermal poly(glutamic acid, threonine) and thermal Poly(glutamic acid, leucine) are the most active of these tested [20]. The activity is gradually decreased by progressive acid hydrolysis [20]. Compared with natural enzymes, the activity of proteinoid is weak. However the decarboxylation of pyruvic acid by proteinoid obeys Michaelis-Menten kinetics as expressed by the Lineweaver-Burk plot [20]. In this reaction a small amount of acetaldehyde and acetoin are formed in addition to acetic acid and CO_2 [20].

3.3.3 Glucuronic Acid

Proteinoids or proteinoid microspheres catalyse the decomposition of D-glucose in aqueous solution to produce CO_2 [21]. The activity observed concerns the decomposition of uniformly labelled ^{14}C-glucose to $^{14}CO_2$ at a low rate of conversion. Glucose labelled only in the 6-position shows a comparable rate of production of $^{14}CO_2$, wheareas glucose labelled in the 1- or 2-position yields virtually no $^{14}CO_2$ [21]. One of the intermediates is glucuronic acid. Glucuronic acid itself has been found to be decarboxylated in the presence of proteinoids [21]. Fructose or sucrose also yields CO_2, but in smaller proportion than does glucose. Some increase in rate of evolution of CO_2 is observed when ATP or Mg^{2+} or both are added to the reaction mixture [21].

In a typical experiment, 100 mg of proteinoid dissolved in 50 ml of water is incubated at 37.5 °C for 3 days with 63 μg of ^{14}C-glucose. The evolved $^{14}CO_2$ is trapped as barium carbonate [21]. Lysine-rich proteinoids are more active than acidic proteinoids. The proteinoids are active over the entire range of pH 1–9 tested. The optimum pHs for lysine-rich proteinoid are 2 and 5 [21]. An experiment in which a second charge of substrate is decomposed as rapidly as the first charge represents one experiment to test whether the activity is truly catalytic [21]. The intermediate glucuronic acid extracted from the reaction mixture is identified by paperchromatography and a positive color test with carbazole [21]. The glucose is more than 90% converted to other materials by proteinoids. At least a small fraction of the glucose is first oxidized to glucuronic acid which is then decarboxylated; each of these reactions is catalysed by the proteinoids [21].

3.3.4 Photochemical Decarboxylation

Light enhances decarboxylation activity by proteinoids, with pyruvic acid, glucuronic, acid, glyoxalic acid, citric acid or indole-3-acetic acid as substrates [22, 23]. In a typical experiment, 20 mg of proteinoid is incubated with 20 μmoles of radioactive substrate for 2–3 days in 10 ml of buffer pH 4.5 (or 7.0) at 37 °C, under the irradiation of a tungsten filament bulb with a filter of 10 % $CuSO_4$ solution; the CO_2 evolved is trapped as sodium carbonate [22].

Irradiation with visible light of the samples containing indole-3-acetic acid yields up to fortyfold increased rates over the velocity in darkness [22]. In general, acidic proteinoids are more active than neutral or basic proteinoids for photodecarboxylation [23]. Considerable complexity of the light reaction is indicated; kinetic studies with pyruvic acid or indole-3-acetic acid do not give a straight line for the Lineweaver-Burk plot [22, 23]. The photosensitization is supposedly due to an unidentified yellow pigment, which is firmly bound to the proteinoid molecules [22]. The spontaneous decay rate for pyruvic acid plus free amino acids in light is 3.2×10 mole/min compared to the lowest value in these experiments 50.5×10 [23]. In a comparison of proteinoids with colored proteins or protein products, both ferritin and casein hydrolysate display a sparing effect on the pyruvic acid substrate, while proteinoids give rates substantially greater than those of the control [23].

Decarboxylation of indole-3-acetic acid enhanced by thermal polylysine in darkness (without irradiation of white light) follows Michaelis-Menten kinetics, thus indicating a reversible catalyst-substrate interaction [22].

3.4 Amination and Deamination

3.4.1 Amination

Thermal polylysine catalyzes the formation of glutamic acid from α-ketoglutaric acid with urea and Cu^{2+} [24]. A reaction mixture of $CuSO_4$, urea, α-ketoglutaric acid (0.1 mM each) and 10 mg of proteinoid in 6 ml of pH 7.0 buffer is incubated at 37.5 °C for 24 hours. The glutamic acid formed in the reaction mixture was identified on an amino acid analyzer after desalting on an ion-exchange column [24].

Kinetic studies of this reaction have shown that it obeys Michaelis-Menton kinetics as expressed by the Lineweaver-Burk plot, the Michaelis constant (K_m) for this reaction at pH 7.0 and 37.5 °C being 2.86×10^{-4} M [24]. Free lysine, Leuchs Poly-L-lysine, total hydrolyzates of thermal polylysine, and amino group-modified thermal polylysine are completely inactive. The activity of thermal polylysine depends on the degree of polymerization [24].

3.4.2 Deamination

Thermal polylysine also catalyzes the formation of α-ketoglutaric acid from glutamic acid with CuCl [25]. A reaction mixture of lysine-rich proteinoid (20 mg), $^{14}C(U)$-L-glutamate (0.1 mM), and CuCl (0.1 mM) in 6 ml of Britton-Robinson buffer (pH 7.0) is incubated at 37.5 °C for 2 hours. More than 40 % of the radioactivity used is recorded in α-ketoglutaric acid by paperchromatography of the reaction mixture [25]. Free lysine and Leuchs poly-L-lysine have no activity [25]. The reaction obeys Michaelis-Menten kinetics at optimum pH 7.0 [25].

3.5 Oxidoreduction

Hemoproteinoids have peroxidatic and catalatic activity [26]. The peroxidase activity of hematin is increased up to 50 times when the hematin is incorporated into proteinoids [26]. The hemoproteinoids have been synthesized from various mixtures of amino acids containing 0.25–2.0 % hematin. The isoelectric point of the lysine rich hemoproteinoids is about 8 and the molecular weights are a little below 20,000 by gel filtration [26].

In a typical kinetic test for the peroxidatic activity in the oxidation of guaiacol, a reaction mixture containing 2.89 ml of 0.1 M phosphate buffer, pH 7.0, 0.05 ml of 0.22 % guaiacol, 0.02 ml of 0.5 % proteinoid solution and 0.04 ml of diluted hydrogen peroxide is incubated for about 30 min. After incubation the colored reaction product is measured by 436 nm light [26]. The catalatic activity in the reaction mixture including guaiacol is measured by peroxide consumption through titration with potassium permanganate [26].

The peroxidatic activity of hemoproteinoids, particularly, increase with their lysine content whereas the catalatic activity especially decreases in proteinoid with high phenylalanine content, hemo-polylysine (hematin is heated with Leuchs polylysine) has very weak peroxidatic activity [26]. The relatively broad pH optimum of lysine-rich hemoproteinoid in the guaiacol test is in the neutral range [26]. Nicotinamide adenosine dinucleotide-reduced form, NADH, is also oxidized by the hemoproteinoids [26].

3.6 Metabolic Pathways

A small part of metabolic pathways by proteinoids has been conceptualized [2, 5]. A flow from oxaloacetic acid to pyruvic acid [18, 19] to acetic acid [20] and a side reaction from pyruvic acid to alanine, which is reversible are depicted in a conceptual integration of results [2, 5].

$$\text{Oxaloacetic acid} \xrightarrow[\text{pH 5}]{\text{Basic proteinoid}} \text{Pyruvic acid} \xrightarrow[\text{pH 8.3}]{\text{Acidic proteinoid}} \text{Acetic acid}$$

Basic proteinoid Cu^{2+} Basic proteinoid Cu^+

Alanine

Each of the reactions is catalyzed by a different type of proteinoid or metalproteinoid complex. The reaction of pyruvic acid to alanine and the reverse reaction are hypothesized from the experimental results in the amination of α-ketoglutaric acid [24] and the deamination of glutamic acid [25].

3.7 Peptide Synthesis in General

The intrinsic properties of proteinoids suggest a model of the assembly of a self-ordered structure representing the first cell. The more complex syntheses by the microspheres themselves require, in addition to enzymes, the provision of energy donor molecules such as ATP. Both ribosomal and aribosomal synthesis of peptides

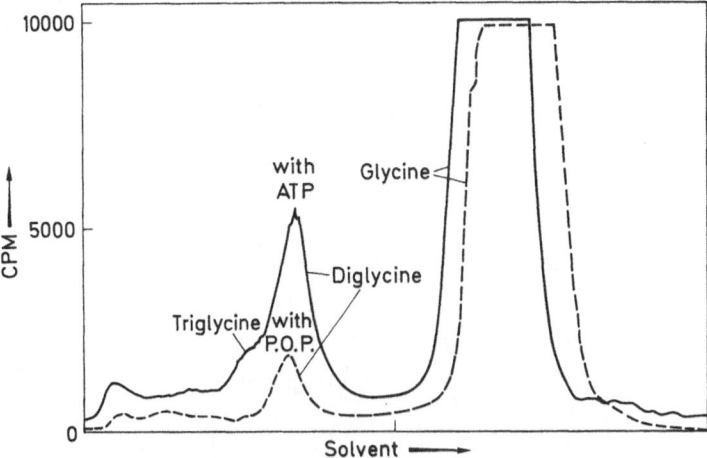

Fig. 3. The synthesis of oligoglycine by proteinoid and ATP (or pyrophosphate): Nakashima, Fox [27]

in modern cells occurs with the aid of ATP. As in the cellular syntheses, which occur at very low concentration, identification and evaluation of yield of peptides require radioactively labelled amino acids for monitoring the syntheses. Problems in the metabolism of proteinoid microspheres require design and interpretation of experiments in a mode of biological chemistry rather than in one of traditional organic chemistry with its emphasis on yields in batch synthesis.

3.7.1 Peptide Synthesis by Proteinoids

Lysine-rich proteinoids in aqueous solution catalyze the formation of peptides from free amino acids, ATP (or pyrophosphate) and Mg^{2+}. Figure 3 shows experiments in which diglycine and triglycine are thus produced [27].

The peptide synthesis is carried out with 200 mg of ATP trisodium salt, 200 mg of radioactive glycine (100 μCi), 200 mg of proteinoid, and 80 mg of $MgCl_2$. These are made up to 1.0 ml in water and the pH is adjusted with concentrated NaOH. After incubation at 25 °C for several days, an aliquot of the reaction mixture is paper-chromatographed, and the radioactivity of peptide separated on the paper is counted on a radiochromatogram scanner [27]. The identity of glycylglycine formed is confirmed also through high voltage electrophoresis, through hydrolysis to glycine only, by study of the kinetics which showed the synthesis to be second-order reaction based on glycine, through chromatography of the dansylated glycylglycine, and through hydrolysis of the dansylglysylglycine to dansylglycine + glycine [27].

This catalytic activity is not found in acidic proteinoids nor in neutral proteinoids, even though they contain some basic amino acid, only in basic amino acid-rich proteinoids. No peptide is formed in controls containing free amino acids. The pH optimus for the synthesis is about 11, but is appreciable below 8 and above 13, the temperature data indicate an optimum at 20 °C or above, with little increase in rate to 60 °C [27]. The yield from this experiment, based on glycine, is 0.40% for diglycine and 0.12% for triglycine [27].

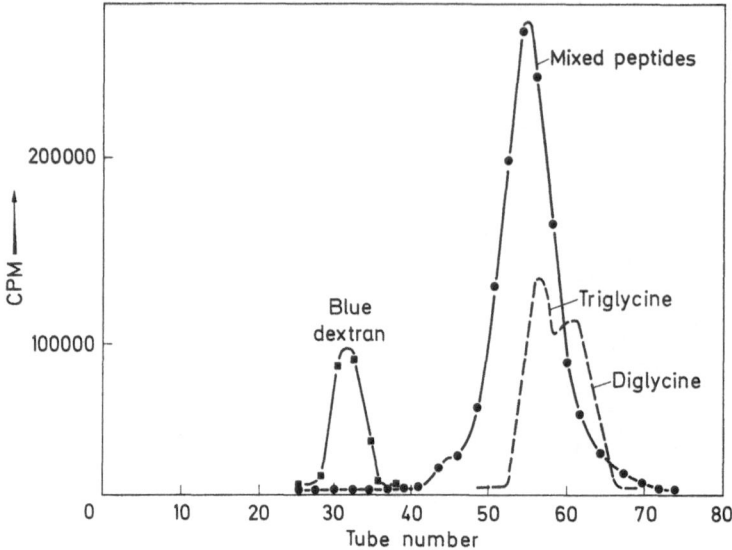

Fig. 4. Fractionation on Sephadex G-25 of the peptides of eighteen amino acids in the presence of histidine, lysine-rich proteinoid and ATP at pH 11 for 5 days: Fox, Nakashima [29]

While the experiments of Fig. 3 were performed with glycine, other peptides have been synthesized from other amino acids such as lysine, phenylalanine or proline [27, 28]. Peptide synthesis from an eighteen amino acid mixture has been also demonstrated by using (histidine and lysine)-rich proteinoid. When the product is fractionated on Sephadex G-25, most of the oligopeptides appear to be in the dipeptide-tripeptide range or larger, and little of free amino acids survive from the reaction. Virtually all types of amino acid appear to yield peptides (Fig. 4) [29].

The synthesis of peptide has been tested at pH 7.2 in a suspension of (acidic + basic) proteinoid microspheres. The activity of the complex particles is several times as large as that of the basic proteinoid solution alone [28].

The syntheses of peptides are energized by ATP, they are operative in the presence of water, and the ATP is most effective in the peptide synthesis when supplied continuously. If the ATP is supplied in repeated small fractions, the conversion from amino acid to peptide in the reaction mixture with acidic and basic proteinoid microspheres is greater than if the same total amount is supplied at once, since the amino acid is present in considerable molar excess over ATP. The most likely explanation for the higher rate of conversion for repeated reaction of ATP is that it is more efficiently used, due to rapid decay of any one charge of ATP [28].

Lysine-rich proteinoid can synthesize peptides with inorganic pyrophosphate instead of ATP (Fig. 3) [27]. Pyrophosphate has been proposed as an evolutionary precursor of ATP [30]. In some microorganisms, pyrophosphate serves as a source of energy in fermentation. The experiment of Fig. 3 compares equimolar amounts of these two energy-rich phosphates. The ratio of yield from ATP to that from pyrophosphate in one experiment was 3.7 [27].

3.7.2 Amino Acid Activation

Nonenzymic transphosphorylation catalyzed by divalent metal ions may have been the prototype reaction for the biochemical evolution of those enzymes which utilise the free energy of hydrolysis of ATP to drive chemical transformation.

Glycyl hydroxamate as a model for biological activation of carboxyl has been synthesized from glycine and hydroxylamine by ATP and divalent cation [31-33]. In these experiments, the yield of the glycyl hydroxamate at 37 °C for 13–48 hours is approximately 0.1 % based on glycine, and the optimum pH is 4 to 5. ATP is far more active than other nucleoside triphosphates [32]. The initial product is probably an acylphosphate or an acyl derivative of AMP (aminoacyl adenylate) [31]. Little or no acylphosphate accumulates when NH_2OH is omitted. This is probably due to the unfavorable equilibrium of acylphosphate formation, and in the presence of hydroxylamine the equilibrium is displaced in favour of hydroxamate formation [31].

The formation of glycyl hydroxamate from the mixture of glycine, hydroxylamine, ATP and Mg^{2+} has been stimulated by proteinoid microspheres prepared from acidic proteinoid, basic proteinoid, and calcium [32]. A reaction mixture of 0.012 M ^{14}C-glycine, 0.05 M hydroxylamine, 0.05 M ATP, 0.05 M $MgCl_2$ and microspheres is incubated at pH 5.0 or 7.0 at 37 °C for 2 days. The product, ^{14}C-glycyl hydroxamate, is separated by high voltage paper electrophoresis and measured on a radiochromatogram scanner [32]. In these experiments, larger effects of proteinoid microspheres have been found at pH 7.0 than at pH 5.0. The yield of glycyl hydroxamate at pH 7.0 with the proteinoid microspheres is three times as much as in the control containing no microspheres. This enhancement is very likely due to the particulate structures, but whether the reaction occurs in or on the microspheres, and the actual role of each proteinoid component, has not yet been determined [32].

It is remarkable that proteinoid microspheres bring the optimum pH for the amino acid activation from the acidic range to neutral. Acidic condition for the activation may be provided in the metal-proteinoid microspheres in neutral buffer.

3.7.3 Peptide Synthesis by Imidazole

It was expected that if amino acids were used instead of hydroxylamine, peptides would be formed in the experiments of aminoacyl hydroxamate formation [31]. However, when glycine is warmed with ATP and $MgCl_2$ in the solid state (derived from the solution), peptide is not formed but adenylyl-(5' → N)-glycine (gly-N-pA) is formed [34, 35]. When imidazole is used instead of amino acids, adenosine 5'-phosphorimidazolide (ImpA) is formed in a few percent yield [34].

Synthesis of peptides from amino acids in the solid state or in aqueous solution has been reported with the aid of ATP, Mg^{2+} and free imidazole [34-36]. A 4.5% yield of glycine peptides has been reported in the solid state, 0.6% yield in aqueous solution [36]. Gly-N-pA is formed in a yield of 72.5% from glycine and chemically synthesized ImpA in aqueous solution, pH 8.0, at room temperature [37]. At initial of pH 6.0 glycyl 5'-adenylate (gly-O-pA) and 2'(3')-O-glycyl adenosine 5'-phosphate (pA-gly) are formed [37].

3.7.4 Amino Acid Derivatives

Amino acyl 5'-adenylates have been known as products in the first stage of activation for the amino acids in protein biosynthesis before they are coupled to t-RNA. The amino acyl 5'-adenylates can be chemically synthesized, when aminoacid, AMP, and dicyclohexylcarbodiimide, as condensing agent, are mixed in aqueous pyridine at 0 °C [38].

The aminoacyl adenylates react rapidly with amino acids to yield peptides [39], under physiological conditions in aqueous solution at room temperature. At pHs higher than 7, the maximum is attained at pH 10 where the extent of the polymerization of peptides from alanyl adenylate is about 60% [40]. The aminoacyl adenylate in aqueous solution undergoes an intra- or inter-molecular rearrangement with the formation of 2'(3')-aminoacyl ester of adenosine [40].

Basic proteins, histones, increase the rate of peptide formation from aminoacyl 5'-adenylates in neutral solution [41]. The acceleration is about threefold in the initial phase of the reaction. A mixture of histone and polyribonucleotide also increases the rate of peptide formation [41]. Histone might offer local basic surroundings to the reacting aminoacyl adenylate molecules [41].

N-Acetylglycyl adenylate has been readily converted in high yield to N-acetylglycyl imidazolide in the presence of excess imidazole at pH 7.0 at room temperature. The aminoacyl group can then be transferred from imidazole to become an ester of the nucleotide [42]. N-Acetylglycyl imidazolide is an activated intermediate in the transfer process. The imidazolide is very unstable at low pHs and high pHs, the most stable region being about pH 7.0, the pK_a of imidazole [42]. The imidazole does not catalyze the decomposition of the ester, this in turn indicates that the imidazolide does not form from the ester [42]. The aminoacyl imidazolides and aminoacyl 5'-adenylates each react specifically with hydroxyl groups in preference to water. Thus, in aqueous solution a proportion of 2'(3')-ester is formed instead of reacting directly to form peptides [33].

Tripeptides or larger peptides are formed from chemically prepared N-(glycyl)-imidazole in aqueous solution in 8 hours [43].

3.7.5 The Mechanism of Peptide Synthesis by Proteinoids

The mode of peptide synthesis by proteinoids or proteinoid microspheres has not so far been ascertained. We propose here possible ways to synthesize peptides from amino acid and ATP by proteinoids, as shown below. Since it is difficult to deduce mechanisms from the data reported, the following scheme is tentative.

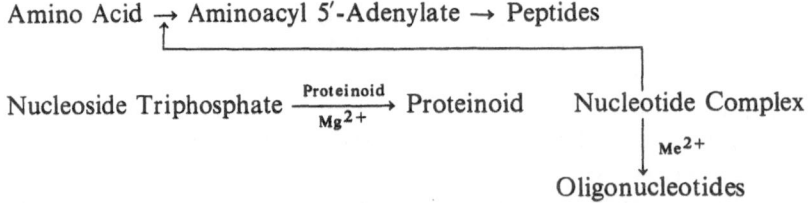

The mechanisms we propose involve proteinoid as a catalyst. The reaction sequences are mostly adopted from the imidazole-catalyzed peptide syntheses reviewed in the previous section.

The intermediate for peptide synthesis is probably aminoacyl 5'-adenylate, formed from amino acids and proteinoid nucleotide complex. In the proteinoid nucleotide complex, the phosphate of adenylate may be attached to the imidazole of histidine in the proteinoid. The aminoacyl 5'-adenylate may react with amino acids to yield peptides through amino acid adenylate ester. There are many alternative ways conceivable according to the experiments reviewed in the previous sections. For example, proteinoid nucleotide complex may be formed through amino acid adenylate phosphoramidate, or aminoacyl adenylate may be formed through amino acid proteinoid complex.

Interaction between proteinoid (made from Lys, His, and Arg) and AMP (or ATP) has been observed after incubation in Mg^{2+} containing Tris buffer, pH 7.2, at 37 °C for 30 min [44]. The larger sized complex was proved by gel filtration. Phosphoric groups may contribute to the association according to the evidence that adenosine is unable to form complex [44].

Although lysine-rich proteinoid has received the most attention in these experiments, the use of proteinoid sufficiently rich in any basic amino acid is the common denominator of compositional requirement for catalytic activity. While acidic proteinoids moderately rich in basic amino acids are inactive in producing peptides, basic proteinoid rich in either histidine or lysine have been found to be active [27-29]. The contribution to peptide bond synthesis of basic proteinoids is consistent with the knowledge that transpeptidation by ribosomes depends upon groups having a pK_a of 7.5–8.0 (imidazole or N-terminal α-amino) and perhaps one of pK_a 9.4 due to ε-amino of lysine [45]. Although the catalysis of peptide bond formation by the ribosomes of contemporary organisms is the subject of many studies, so far it has not been possible to understand completely the mechanism of peptide bond formation. Two hypotheses have been proposed [46]. One postulates nucleophilic or acid-base catalysis by the functional groups of the peptidyltransferase center. According to the other hypothesis peptidyltransferase catalyzes peptide bond formation merely by sterical orientation of the reaction components. The peptide bond is synthesized spontaneously [46]. The catalytic mechanism of peptide bond formation in ribosomes remains obscure [46].

Acidic proteinoid potentiates the active structure of lysine-rich proteinoid participating in forming microspheres in neutral buffer. Physical surface effects and providing micro condition in the microspheres could be surmised. Activation of amino acids generally requires acidic condition. Amino acids are activated by ATP and Mg^{2+} at pH 4–5 [32, 33]. Aminoacyl adenylate anhydride and ester is formed preferentially from amino acid and adenylate imidazolide at pH 6.0 [37]. On the other hand, polycondensation of activated amino acids undergoes at pH values higher than 7. Peptides are formed from aminoacyl adenylate in basic buffer (the optimum pH is 10 for alanyl adenylate; [40] from amino acid adenylate phosphoramidate and imidazole at pH 7.0 [34], from N-(aminoacyl)-imidazole at pH 6–9 [43]. In this context, acidic and basic environments may be provided inside and/or on the surface of the microspheres composed of acidic and basic proteinoids in neutral buffer. Acidic micro condition suitable for the activation of amino acids and basic micro environment favorable for peptide formation from activated amino acid may be provided.

3.8 Internucleotide Synthesis

Some proteinoids catalyze synthesis of internucleotide bonds. Adeninedinucleotide is formed from the action of lysine-rich proteinoid with ATP and $MgCl_2$, while a

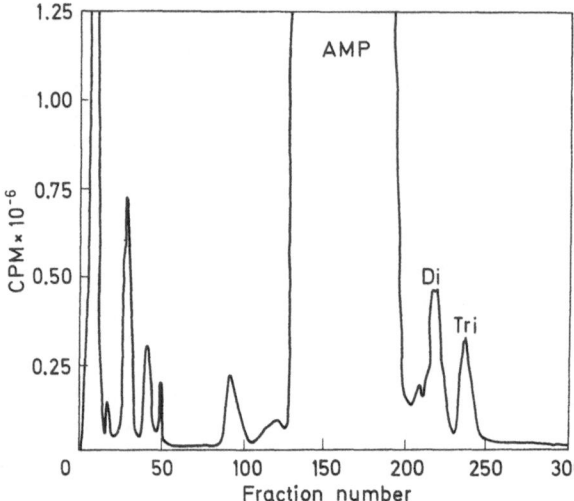

Fig. 5. The synthesis of oligo-nucleotides by proteinoid and ATP: Jungck, Fox [47]

suspension of microspheres of acidic and basic proteinoids is used, dinucleotide and trinucleotide of adenine are produced (Fig. 5) [47].

A mixture of 0.44 g of acidic proteinoid, 0.19 g of basic proteinoid, and 0.31 g of radioactive ATP in 2.0 ml of 0.02 M $MgCl_2$, 0.05 M Tris buffer, is heated in a boiling water bath for 5 min, and then incubated at 37 °C for 24 hrs, after the fractionation of the mixture by DEAE-cellulose column chromatography. Oligonucleotides produced are identified with those of authentic markers [47]. Chain length has been determined by the ratio of AMP to adenosine in the alkaline hydrolyzate of the material which was treated by alkaline phosphatase to remove 5'- and 3'-phosphate [47].

ATP is converted to dinucleotide in magnesium-containing buffer without proteinoid in a yield of 0.7%, but in the presence of basic proteinoid solution, acidic proteinoid microspheres, or acidic-basic proteinoid microspheres, the conversion to oligo-nucleotides is 2.2–2.3%. With AMP instead of ATP no oligonucleotides result. The proteinoids are thus a model for proto-RNA polymerase activity, and ATP is reaffirm-ed as the source of energy for the synthesis of phosphodiester linkages [47]. The micro-spheres of acidic proteinoid or of acidic-basic proteinoid synthesize di- and tri-nucleotides, while without proteinoid or basic proteinoid solution dinucleotide only results. Furthermore, the ratios of trinucleotides/dinucleotide are 0.5 for acidic-basic proteinoid microspheres, 0.2 for acidic proteinoid microspheres [47].

Polymerization activity of proteinoid or polynucleotide-phosphorylase has been compared in ADP solution at pH 8.5 [48]. The results show that the activity of the neutral proteinoid is approximately 20 times lower than that of the enzyme, and the lower molecular-weight fraction of the proteinoid has negligible activity. The polymeriza-tion by polynucleotide-phosphorylase is increased approximately three times when Leuchs polylysine is supplied, polylysine and enzyme yield a heterogeneous solution [48].

Simultaneous peptide and oligonucleotide formation has been examined on a mixture of amino acid, nucleoside triphosphate, imidazole, and $MgCl_2$ as reviewed in the previous section. In this reaction the oligonucleotide probably forms via nucleoside 5'-phosphorimidazolide [36]. The yield of dinucleotide from ATP is up to 0.12%, the product is ApAp, ApApp or ApAppp [36]. ApA, which is formed from the product by phosphatase hydrolysis, is further characterized by digestion with snake venom phosphodiesterase to yield 50% adenine and 50% 5'-AMP [36].

The mechanisms of nucleotide oligomer synthesis catalyzed by proteinoids have not reported. A tentative mechanism rests on the supposition that the proteinoid nucleotide complex is an intermediate of an enzyme-like nucleotidyl transfer reaction.

$$\text{Nucleoside Triphosphate} \xrightarrow[\text{Mg}^{2+}]{\text{Proteinoid}} \text{Proteinoid Nucleotide Complex}$$
$$\downarrow \text{Me}^{2+}$$
$$\text{Oligonucleotides}$$

A histidine residue localized in the active site of the proteinoid may be the primary acceptor of the nucleotide. The nucleotide residue bound to the proteinoid would then transfer to the final acceptor. The metal ion may work as a transfer catalyst.

The following related reports may be useful to understanding the mechanisms of proteinoid-catalyzed oligonucleotide synthesis, and to further development of the study. When 2'-amino-2'-deoxyuridine reacts with nucleoside 5'-phosphorimidazolides in aqueous solution, the dinucleotide monophosphate analogues are obtained in a yield exceeding 80% [49]. A divalent ion catalyzes the condensation of nucleoside 5'-phosphorimidazolides in aqueous solution, pH 7.0, at 0 °C, in 7 or 10 days, to form oligonucleotides in yields of up to 58% [50]. Among the divalent ions, Pb^{2+}, Zn^{2+}, Mn^{2+} etc., Pb^{2+} is most effective, and about 90% of the di- and tri-nucleotides produced have 2'-5'-phosphodiester linkage [50]. Poly (U) template-directed condensation of adenosine 5'-phosphorimidazolides catalyzed by the Pb^{2+} ion produces predominantly 3'-5'-linked (75%) oligoadenylates [51]. Oligoadenylates synthesized from guanosine 5'-phosphorimidazolide on an oligocytidylate template in the presence of Zn^{2+} have predominantly 3'-5'-linkage [52].

The facts that basic proteinoids catalyze the synthesis of both peptide and internucleotide bonds and that they do this in suspensions of microspheres, i.e. in the same microlocale has led to the suggestion that microspheres on the primitive Earth were the site of development of the coded genetic mechanism [53].

3.9 Nucleoproteinoid Microspheres

3.9.1 Microsystems from Basic proteinoid and polynucleotide (Nucleoproteinoids)

Experiments have revealed that the microsystems of proteinoid and polynucleotide have stability in solution under changing temperature of pH such as is not possessed by particles composed of acidic proteinoid alone [2, 54]. The nucleoproteinoid microparticles have been viewed as models of evolutionary precursor of ribosomes [54].

Various lysine-rich proteinoids were tested for their ability to form microparticles with yeast RNA. This ability was found to appear at a proportion of basic to dicarboxylic amino acid of approximately 1.0 [54]. Mixing of dilute solutions of lysine-rich proteinoid and of RNA yields small globular microparticles which are not dissolved by heating. On the other hand, when a sufficiently lysine-rich proteinoid is allowed to interact with calf thymus DNA, fibrous material results [54].

Markedly different morphologies resulted when acidic proteinoid was combined with basic polymer in the presence of calcium. The microspheres obtained by combining calf thymus histone with acidic proteinoid are highly complex in structure [55]. When thermal histonoid (proteinoids having histone-like composition) has been substituted for the thymus histone, these microspheres are virtually indistinguishable in morphology from those made with the organismic histone [55]. Either polynucleotide or acidic proteinoid can serve as polycation in polycationic-polyanionic microspheres [3].

Effect of Mg^{2+} on the formation of nucleoproteinoid microparticles has been studied, because of its biological significance, for example, in maintaining ribosomal stability [56]. Mg^{2+} has a marked effect on the amount of the nucleoproteinoid particles, the precipitate of nucleoproteinoid particles decreasing as the Mg^{2+} concentration increases [56]. The effects with Mg^{2+} are perhaps explained as a competitive binding of the Mg^{2+} to the polynucleotide. Mg^{2+} tends to displace lysine-rich proteinoid from the particles [56].

The amount of material precipitated tends to decrease (except with poly G) as the lysine content in the proteinoid increases. This phenomenon is visualized as due to reduction in intraproteinoid sidechain interaction and outlooping [56].

3.9.2 Selective Formation

Lysine-rich proteinoids preferentially interact to form microparticles with polycytidylic acid and polyuridylic acid. On the other hand arginine-rich proteinoids interacts much more readily with polyguanylic acid and polyinosinic acid to form such particles under the same condition [57]. Lysine-rich proteinoids had most often shown a preferential precipitation with polyuridylic acid in the experiments for interaction between polynucleotides and the proteinoids in the presence of Mg^{2+} [56].

Selective formation of microparticles from polynucleotides and lysine-rich proteinoids rich in individual radioactive amino acid has been studied and the focus of attention is on those homoanticodonic amino acids having one homogeneous codon (glycine, CCC; lysine, UUU; proline, GGG; and phenylalanine, AAA) [58]. Precipitation of individual amino acid rich proteinoids with each of homopolyribonucleotides, with and without Mg^{2+}, was tested [58]. The results show that three (Lys-rich, Gly-rich,

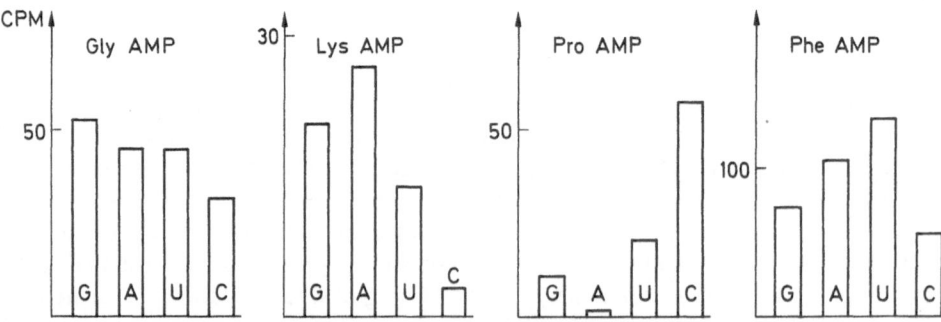

Fig. 6. Relative effects of nucleoproteinoid microspheres and individual homopoly-nucleotides on condensation of each of four aminoacyl adenylates: Nakashima, Fox [59]

Table 1. Analysis of peptides synthesized, Nakashima, Fox [28]

Catalytic agent	Homopeptide	Heteropeptide	Heteropeptide	
			Gly-Phe	Phe-Gly
Imidazole (Proteinoid microspheres)	75%	25%		
Acidic-basic	32	68		
Poly(A)-basic	30	68	82%	18%
Poly(G)-basic	42			
Poly(c)-basic	37	58	24	76
Poly(U)-basic		64	32	68

and Pro-rich) of the four proteinoids studied yielded results consistent with a matrix of anticodonicity; the fourth (Phe-rich) results were consistent with matrices of codonicity and anticodonicity.

3.9.3 Selective Condensation of Amino Acids

Microparticles composed of each of four homopolyribonucleotides and the same lysine-rich proteinoid is found to influence the incorporation of individual aminoacyl adenylate [59]. The incorporation favors the amino acids whose codons are related to the nucleotide in the particles (Fig. 6), when conditions are appropriately chosen. Other conditions yield other preferences. These results support a stereochemical basis for the genetic code.

Formation of oligopeptides of each of four amino acids in suspension of nucleoproteinoid microparticles composed of polyadenylic acid and lysine-rich proteinoid has been studied. Amino acids are each converted to peptides by the nucleoproteinoid microparticles in suspension, at rates that differ in minor degree for the various amino acids [28].

When equimolar solutions of mixtures of glycine and phenylalanine with ATP are tested in suspensions of nucleoproteinoid microparticles of lysine-rich proteinoid and each of various polyribonucleotides, both homopeptides and heteropeptides are synthesized [28]. Glycylphenylalanine or phenylalanylglycine is the principal product; the preference is related to which polyribonucleotide is in the complex (Table 1). The effect of imidazole substituted for the nucleoproteinoid microarticles on glycine and phenylalanine is the production predominantly of the homopeptides diglycine and diphenylalanine, with 25% of a mixed heteropeptide fraction. In the other cases, acidic-basic proteinoid microspheres and nucleoproteinoid microspheres, the heteropeptides are the main product and the homopeptides represent a minor fraction. These nucleoproteinoid microspheres indicate a decided preference for synthesis of the gly-phe type when the polynucleotide is poly A. With poly C or poly U, the main product in the heteropeptide fraction is the obverse one of phe-gly.

The results do display a parallelism in reaction for the two pyrimidine polynucleotides. This parallelism is to that described earlier in static interactions of arginine-rich and lysine-rich proteinoids with polynucleotides [57]. This parallelism suggests for the dipeptides a synthetic mechanism in which one kind of amino acid is bound by the polynucleotide while the other amino acid is reacted with the bound one.

3.9.4 Model Ribosomes

In exhibiting the ability to synthesize peptide bonds and to influence amino acid sequences, the ribonucleoproteinoid microspheres are a model for ribosomes [60, 61]. A kind of acidic proteinoid has three major fractions, of which the molecular weights by the sedimentation analysis are 4070, 5160, and 5800 [62]. It is reasonable to assume that the basic proteinoid is a mixture of fractions having consecutive molecular masses since the basic proteinoid is an analog of acidic and neutral proteinoids [63]. The results of fractionation of tritonic proteinoid (Glu, Gly, and Tyr) show consecutive glycine-containing peptides [4]. The proteins from each subunit of ribosomes form a graded sequence of closely related molecular weights within a certain range, and the grading of the protein sizes may be an essential requirement for ribosome function [61]. If this hypothesis is acceptable, the proteinoids are suitable models for ribosomal proteins as well as protobiological models. Homologous tertiary structures between ribosomal proteinas would not necessarily relate to their primary sequences [64].

4 Hormonal and Inhibitory Activities of Proteinoids

4.1 Hormonal Activity

Thermal poly(Arg, Glu, Gly, His, Phe, Trp) or poly(Arg, Glu, Gly, His, Phe) has melanocyte-stimulating hormonal (MSH) activity. The activities are 2.2×10^4 units/grame and 1.4×10^3 respectively, whereas the "active site" of the MSH consisting of six amino acids in one chain shows 2.2×10 [65]. The activity of natural hormone is 3.3×10^5 [65]. Thermal poly(Glu, Gly, His, Phe, Trp) (minus Arg) or poly(Ala, Asp, Glu, Leu, Lys, Phe) has no activity [65].

4.2 Inhibitions

4.2.1 Glyoxalase I

Variations in the composition of proteinoid inhibitory for glyoxalase I, an enzyme which has occupied a central position in Szent-Gyorgyi's theory of tumourgenesis have been studied [66, 67]. The activities of several amino acids in four groups of polymers in which two, three, four, or five amino acids have been heated in various molar ratios have been compared. Those produced from both tryptophan and cysteine are the most active, while the control mixtures of the corresponding amino acids have no activity [66, 67]. Hydrophobicity appears to be a physical property contributing to the activity of some of the inhibitors, the more active fraction of poly(Asp, Glu, Trp, CySH) upon fractionation by dialysis is of higher molecular weight [66]. A Lineweaver-Burk plot of the effect of copoly(Glu, Trp) suggested non-competitive inhibition over one range of concentration. An Eadie-Hofstee plot of the effect of poly(Asp, Glu, Trp, CySH) indicates competitive inhibition of that polymer with glutathione [66].

4.2.2 Other Inhibitions

Lysine-rich proteinoids (thermal polylysine) inhibit the growth of roots and shoots of plants, *Digitalis purpurea* and *Phacelia tanacetifolia* [68]. The growth inhibition by the proteinoid is believed to be due to the decarboxylation of indole-3-acetic acid catalyzed by the proteinoid [22].

A number of proteinoids have been found, in unpublished studies, to be inhibitors of the enzymes: sperm hyaluronidase, pork ATPase, luciferase, and α-chymotrypsin [66].

5 Conclusions and Prospects

The catalytic, synthetic, hormonal, and inhibitory activities that have been found in proteinoids or proteinoid microspheres are listed in Table 2. The possibility that metabolic activities found were due to contamination by micro-organisms is denied by experiments under aseptic condition or by the several experimental observations. The activities of proteinoids are generally weak. In some cases, proteinoids act several orders of magnitude more slowly than do modern enzymes or organisms, but free amino acids or Leuchs polypeptides usually have no activity or less than the proteinoid composed of the same amino acids. In general, activities of proteinoids increase approximately in proportion to its molecular weight. One or more of the proteinoids has been found to meet the salient requirements of enzymes such as Michaelis-Menten kinetics, pH-activity curves, etc.

The weakness of proteinoid function is plausible from a viewpoint of molecular evolution, and from a chemical standpoint it is due to the primitive structure of proteinoid showing a relatively simple, or an elementary, reaction such as the active site of enzyme. For example, the melanocyte stimulating activity of the proteinoid is close to the activity of the active site peptide of the hormone, MSH.

Almost all fractions separated have the activity [9, 66]. In some studies the activity of fractions separated by gel filtration is increased up to about 20 times, but unidentified inhibitor may have been removed during the fractionation [9]. The activity could be improved by the provision of a new active site and structure which could be formed from synthesized polypeptides in proteinoid microspheres. Mixed aminoacyl adenylates have been reacted with proteinoids to yield polymers of substantially increased size, and the modified proteinoids form microspheres [69].

The results to date suggest that the more basic proteinoids are more often catalytic and synthetic than are the acidic and neutral proteinoids. In general, the activity of proteinoid is mostly related to histidine or lysine as a component of proteinoids. Histidine is an imidazole derivative that plays a special role in contemporary biochemistry, as an amino acid present at the active site of a variety of enzymes.

The inactivity of acidic proteinoid (A) synthesize peptides in solution of pH 7.2 is in contrast to the activity of the lysine-rich proteinoid (B), when A and B form microspheres. The activity of the complex is several times as large as that of lysine-rich proteinoid alone [28]. Physical surface and interior providing suitable environment for dehydration may be responsible. It may also be that the closed molecules of both proteinoids may be opened by the interaction, consequently buried active sites of proteinoids become effective. Neutral amino acids contained in each proteinoid may prevent the entire neutralization of acidic and basic residues. In the synthetic pathways of peptides and oligonucleotides by proteinoid microspheres, proteinoid nucleotide complex is proposed in this article as a main intermediate of the reactions. Oligo-

Table 2. Metabolic Activities of Proteinoids

(Activity)	Proteinoids	Essential	pH	Ref.
(Hydrolysis)				
p-Nitrophenyl acetate	a	His, Asp-imide	6.7	7, 8, 19)
p-Nitrophenyl phosphate	(a), n, b		6.7	9)
3-O-Methyl fluorescein				9)
ATP	Zn-A			11, 12)
(Decarboxylation)				
Oxaloacetic acid	b	Lys-amino	5 or 6	18, 19)
Pyruvic acid	a, (b)	Asp-imide	8.3	20)
	a	light	3.5 or 7.1	23)
Glucuronic acid	(a), (A), b		2 or 5	21)
	a, b	light	8	23)
Indole-3-acetic acid	a, b		4.5	22)
	a, b	light	4.5	22)
Glyoxalic acid	a	light	7.1	23)
Citric acid	a	light	7.1	23)
(Amination)				
α-ketoglutaric acid	Poly-Lys	Lys-amino, Cu^{2+}	7.0	24)
(Deamination)				
Glutamic acid	Poly-Lys	Cu^+	7.0	25)
(Oxidoreduction)				
(Peroxidatic)				
Guaiacol	Hemo-b, (-n), (-a)		7.0	26)
NADH	Hemo-b, (-n), (-a)			26)
(Catalatic)				
H_2O	Hemo-b, (-n), (-a)		7.0	26)
(Synthetic)				
ATP	A	Tyr, Fe^{2+}, light	2–3	13, 14)
Glycine hydroxamate	Ca-AB	ATP, Mg^{2+}	5 or 7	31, 32, 33)
Peptide	b, AB	ATP, Mg^{2+}	11.7	27, 28, 29)
Oligonucleotide	A, b, AB	ATP, Mg^{2+}	7.0	47, 48)
(Hormonal)				
Melanocyte stimulating		Arg, Glu, Gly, His, Phe		65)
(Inhibition)				
Glyoxalase I		Trp, CySH		66, 67)
Indol-3-acetic acid	Poly-Lys			22, 68)
Sperm hyaluronidase				66)
Luciferase				66)
α-Chymotrypsin				66)

a, n, b acidic, neutral, basic proteinoid solution;
A, B acidic, basic proteinoid microspheres;
AB acidic-basic proteinoid microspheres;
( ) proteinoid having weaker activity

nucleotides might form from the intermediate while amino acids may be activated by being in the complex.

Nucleoproteinoids composed from polynucleotide and basic proteinoid have activity to synthesize peptides selectively [28, 59]. The mechanism is unknown. This is a way to resolve the genetic coding mechanism. Affinity, for example, hydrophobicity or hydrophilicity, between amino acid and nucleic acid is related to their anticodonic interaction [58]. Each nucleotide has distinctive affinity with each amino acid. Nucleotides influence the amino acid sequence of peptide. In this context amino acids of proteinoid may influence the nucleotide sequence. Proteinoids and Leuchs peptides influence the polymerization reaction of nucleoside 5'-phosphorimidazolides [70]. This kind of experiment has been done on the theory that proteinoids are the first informational macromolecules in the evolutionary sequence [2]. Experiments for the influence of nucleotide sequence by proteinoids are being tested. Proteinoids synthesize di- and tri-ribonucleotides, but synthesis of larger polynucleotides has not yet been accomplished or shown to be accomplished. Enzymically synthesized polyribonucleotides have been therefore employed for the studies of nucleoproteinoid microspheres. The syntheses of large peptides and nucleotides are wanted. Reciprocal template action between two kinds of oligonucleotide may be a possible way to enlarge each other.

Substantial enzymic and synthetic activities of proteinoid microspheres have been found, in spite of a requirement for a narrowed range of condition and materials for the experiments. These findings are evidence to support the evolutionary theory for proteinoid microspheres.

6 Acknowledgement

I wish to express my sincere thanks to professor S. W. Fox for his criticism of the manuscript. The main bulk of experiments from this laboratory were funded by NGR-10-007-008 of the National Aeronautics and Space Administration.

7 References

1. Fox, S. W., Harada, K.: J. Am. Chem. Soc. *82*, 3745 (1960)
2. Fox, S. W., Dose, K.: Molecular Evolution and the Origin of Life, New York, Dekker 1977
3. Fox, S. W.: Naturwissenschaften *67*, 378 (1980)
4. Nakashima, T., Jungck, J. R., Fox, S. W., Lederer, E., Das, B. C.: International Journal of Quantum Chemistry: Quantum Biology Symposium *4*, 65 (1977)
5. Rohlfing, D. L., Fox, S. W.: Adv. Catalysis *20*, 373 (1969)
6. Dose, K.: The evolution of individuality at the molecular and protocellular levels, in: Individuality and Determinism (ed.) Fox, S. W., p. 33, New York and London, Plenum Press 1984
7. Rohlfing, D. L., Fox, S. W.: Arch. Biochem. Biophys. *118*, 122 (1967)
8. Rohlfing, D. L., Fox, S. W.: Arch. Biochem. Biophys. *118*, 127 (1967)
9. Oshima, T.: Arch. Biochem. Biophys. *126*, 478 (1968)
10. Tetas, M., Lowenstein, J. M.: Biochemistry *2*, 350 (1963)
11. Fox, S. W., Wiggert, E., Joseph, D.: Simulated natural experiments in spontaneous organization of morphological units from proteinoid, in: The Origin of Prebiological Systems (ed.) Fox, S. W., p. 368, New York Academic Press 1965
12. Durant, D., Fox, S. W.: Federation Proc. *25*, 342 (1966)
13. Fox, S. W., Adachi, T., Stillwell, W.: A quinone-assisted photoformation of energy-rich chemical bonds, in: Solar Energy: International Progress, Vol. 2 (ed.) Veziroglu, T. N., p. 1056, New York, Pergamon Press 1980
14. Fox, S. W., Adachi, T., Stillwell, W., Ishima, Y.: Photochemical synthesis of ATP: Protomembrans and protometabolism, in: Light Transducing Membranes: Structure, Function, Evolution (ed.) Deamer, D. W., p. 61, New York, Academic Press 1978
15. Krasnovsky, A. A.: Origins of Life *7*, 133 (1976)
16. Penningroth, S. M., Olenhik, K., Cheung, A.: J. Biol. Chem. *255*, 9545 (1980)

17. Hammes, G. C.: Proc. Natl. Acad. Sci. *79*, 6881 (1982)
18. Rohlfing, D. L.: Arch. Biochem. Biophys. *118*, 468 (1967)
19. Rohlfing, D. L.: Science *169*, 998 (1970)
20. Hardebeck, H. G., Krampitz, G., Wulf, L.: Arch. Biochem. Biophys. *123*, 72 (1968)
21. Fox, S. W., Krampitz, G.: Nature *203*, 1362 (1964)
22. Hardebeck, H. G., Haas, W.: Z. Naturforsch. *26b*, 862 (1971)
23. Wood, A., Hardebech, H. G.: Light enhanced decarboxylations by Proteinoids, in: Molecular Evolution (eds.) Rohlfing, D. L., Oparin, A. I., p. 233, New York, Plenum Press 1972
24. Krampitz, G., Baars-Diehl, S., Haas, W., Nakashima, T.: Experientia *24*, 140 (1968)
25. Krampitz, G., Haas, W., Baars-Diehl, S.: Naturwissenschaften *55*, 345 (1968)
26. Dose, K., Zaki, L.: Z. Naturforsch. *26b*, 144 (1971)
27. Nakashima, T., Fox, S. W.: J. Mol. Evol. *15*, 161 (1980)
28. Nakashima, T., Fox, S. W.: BioSystems *14*, 151 (1981)
29. Fox, S. W., Nakashima, T.: BioSystems *12*, 155 (1980)
30. Lipmann, F.: Projecting backward from the present stage of evolution of biosynthesis, in: The Origins of Prebiological Systems (ed.) Fox, S. W., p. 259, New York, Academic Press 1964
31. Lowenstein, J. M.: Biochim. Biophys. Acta *28*, 206 (1958)
32. Ryan, J. W. and Fox, S. W.: BioSystems *5*, 115 (1973)
33. Lowenstein, J. M., Schatz, M. N.: J. Biol. Chem. *236*, 305 (1961)
34. Lohrmann, R., Orgel, L. E.: Nature *244*, 418 (1973)
35. Sawai, H., Lohrmann, R., Orgel, L. E.: J. Mol. Evol. *6*, 165 (1975)
36. Weber, A. L., Caroon, J. W., Warden, J. T., Lemmon, R. M., Calvin, M.: BioSystems *8*, 277 (1977)
37. Lohrmann, R., Ranganathan, R., Sawai, H., Orgel, L. E.: J. Mol. Evol. *5*, 57 (1975)
38. Berg, P.: J. Biol. Chem. *233*, 608 (1958)
39. Paecht-Horowitz, M., Katchalsky, A.: Biochim. Biophys. Acta *140*, 14 (1967)
40. Paecht-Horowitz, M., Katchalsky, A.: Biochim. Biophys. Acta *140*, 24 (1967)
41. Ishigami, M., Tonotsuka-Ohta, N., Nagano, K., Kinjo, M.: Origin of Life *10*, 293 (1980)
42. Lacey, J. C., White, W. E.: Biochem. Biophys. Res. Comm. *47*, 565 (1972)
43. Weber, A. L., Lacey, J. C.: Biochim. Biophys. Acta *349*, 226 (1974)
44. Honda, Y., Yamanashi, H., Imahori, K., Yuasa, S.: Affinity of adenylate derivatives for thermal Polyamino acids, in: Origin of Life; Proceedings of the Second ISSOL Meeting, the Fifth ICOL Meeting (ed.) Noda, H., p. 315, Japan, Bus. Cent. Acad. Soc. Japan 1978
45. Harris, R. J., Pestka, S.: Peptide bond formation, in: Molecular Mechanisms of Protein Biosynthesis (eds.) Weissbach, H., Pestka, S., p. 413, New York, Academic Press 1977
46. Krayevsky, A. A., Kukhanova, M. K.: The peptidyltransferase center of ribosomes, in: Progress in Nucleic Acid Research and Molecular Biology, Vol. 23, p. 1, New York, Academic Press 1979
47. Jungck, J. R., Fox, S. W.: Naturwissenschaften *60*, 425 (1973)
48. Liebl, V., Novák, V., Bejšovcová, L., Masinovskij, Z., Oparin, A. I.: Polymerization of radioactive adenosine diphosphate by polynucleotide-phosphorylase or by proteinoids in microsystems, in: Origin of Life; Proceedings of the Second ISSOL Meeting, the Fifth ICOL Meeting (ed.) Noda, H., p. 363, Japan, Bus. Cent. Acad. Soc. Japan 1978
49. Lohrmann, R., Orgel, L. E.: J. Mol. Evol. *7*, 253 (1976)
50. Sawai, H.: J. Am. Chem. Soc. *98*, 7037 (1976)
51. Sleeper, H. L., Lohrmann, R., Orgel, L. E.: J. Mol. Evol. *13*, 203 (1979)
52. Fakhrai, H., van Roode, J. H. G., Orgel, L. E.: J. Mol. Evol. *17*, 295 (1981)
53. Fox, S. W.: A model for protocellular coordination of nucleic acid and protein syntheses, in: Science and Scientists (eds.) Kageyama, M., Nakamure, K., Oshima, T., Uchida, T., p. 39, Tokyo, Japan Scientific Societiers Press 1981
54. Waehneldt, T. V., Fox, S. W.: Biochim. Biophys. Acta *160*, 239 (1968)
55. Miquel, J., Brooke, S., Fox, S. W.: Currents in Modern Biology *3*, 299 (1971)
56. Lacey, J. C., Yuki, A., Fox, S. W.: BioSystems *11*, 1 (1979)
57. Yuki, A., Fox, S. W.: Biochem. Biophys. Res. Comm. *36*, 657 (1969)
58. Lacey, J. C., Stephens, D. P., Fox, S. W.: BioSystems *11*, 9 (1979)
59. Nakashima, T., Fox, S. W.: Proc. Nat. Acad. Sci. USA *69*, 106 (1972)
60. Nakashima, T., Fox, S. W.: Model Ribosomes, in: From Gene to Protein: Translation into Biotechnology, The Fourteenth Miami Winter Symposium (eds.) Ahmad, F., Schultz, E. E., Whelan, W. J.: p. 81, University of Miami, School of Medicine 1982

61. Nakashima, T.: Protoribosomes, in Molecular Evolution and Protobiology (eds.) Matsuno, K., Dose, K., Harada, K., Rohlfing, D. L.: p. 215, New York and London, Plenum Press 1984
62. Fox, S. W., Nakashima, T.: Biochim. Biophys. Acta *140*, 155 (1967)
63. Fox, S. W., Waehneldt, T. V.: Biochim. Biophys. Acta. *160*, 246 (1968)
64. Kurland, C. G.: Aspects of ribosome structure and function, in: Molecular Mechanisms of Protein Biosynthesis (eds.) Weissbach, H., Pestka, S., p. 81, New York, Academic Press 1977
65. Fox, S. W., Wang, C. T.: Science *160*, 547 (1968)
66. Fox, S. W., Syren, R. M., Windsor, C. R.: Thermal copoly(amino acids) as inhibitors of glyoxalase I, in Submolecular Biology and Cancer (ed.) Wostenholme, G., p. 175, London, Ciba Foundation 1979
67. Syren, R. M., Windsor, C. R., Fox, S. W.: International Journal of Quantum Chemistry: Quantum Biology Symposium *6*, 283 (1979)
68. Haas, W., Hardebeck, H. G.: Naturwissenschaften *56*, 220 (1969)
69. Krampitz, G., Fox, S. W.: Proc. Nat. Acad. Soc. USA *62*, 399 (1969)
70. Nakashima, T., Fox, S. W.: in preparation

Organic Matter in Carbonaceous Chondrites

Françoise Mullie and Jacques Reisse*

Université Libre de Bruxelles, Laboratoire de Chimie Organique (CP 165),
50, avenue F. D. Roosevelt, B 1050 Bruxelles, Belgium.

Table of Contents

---

* Author to whom all correspondance should be addressed. This paper is dedicated to Professor E. Picciotto.

Carbonaceous chondrites are objects of great interest for chemists and astrophysicists. They contain a large number of abiogenic organic molecules and they are fragments of the least metamorphized bodies which exist in the solar system. Therefore, the study of carbonaceous chondrites affords information about the state of matter 4.5 Gyr ago. In this context, the presence of organic molecules is of prime interest even if the origin(s) of these molecules remains obscure. Signatures of the synthetical pathways can be found in the structure of these molecules itself but also in the isotopic ratios of the constitutive atoms. Some very abnormal isotopic ratios seem to prove that the protosolar nebula never achieved a state of complete mixing and that intact presolar fragments are trapped in the chondrite matrix.

Nevertheless, all abnormal isotopic ratios are not necessarily related to an initial heterogeneity of the protosolar nebula. Magnetic isotope effects could have played a role during the accretion and even later, during the long life-time of the chondrites parent bodies.

1 Introduction

In this review article we would like to show why the study of the organic matter in carbonaceous chondrites (a class of meteorites) is so interesting. To this end it is necessary to provide information on the nature, origin and age of these extraterrestrial objects.

The organic chemists involved in natural product research know that the real interest of their work lies in the fact that it contributes to a better understanding of living systems. The establishment of an exhaustive list of the secondary metabolites found in plants is not, per se, of major interest and cannot be considered as fundamental research. The same is obviously true for organic chemists involved in the area of natural product chemistry dealing with the study of the organic matter in meteorites. Such a study is interesting if, and only if, it contributes to a better knowledge of the solar system. As we will demonstrate, this better knowledge is intimately related to the origin of our solar system and the state of matter in the protosolar nebula 4.5 billion years ago and, possibly, to the origin of life on earth. Considered in this context, the study of the organic matter in meteorites can be described as fundamental research; it points to the answers to some fundamental questions.

In writing this review article, we have also tried to treat the subject in a different manner from, and, if possible, to complement the excellent review published in 1981 by Hayatsu and Anders in this same series [1].

2 Historical Background

Carbonaceous chondrites are members of a class of stony meteorites called chondrites. As is clearly indicated by their name, carbonaceous chondrites contain carbon. This carbon itself is present as elemental carbon and also as components of organic molecules.

The presence of organic molecules in samples of extraterrestrial matter has been known for more than a century. Some of the greatest chemists of the nineteenth century were involved in the analysis of samples of meteoritic material. They were able to show that carbonaceous chondrites (as they are now named) contain organic molecules. The first to detect carbon in a meteoritic sample was Thenard, in 1806, by analysis of a sample of the Alais meteorite. This result was confirmed in 1834 by Berzelius, who was also the first to detect the presence of water of crystallisation. Working on a sample of the Kaba meteorite, Wöhler (1858) confirmed the presence of organic matter, and in a paper dated 1859 said, "I am still convinced that besides free carbon this meteorite contains a low-melting point, carbon containing substance which seems to be similar to certain fossil hydrocarbon-like substances . . .".

The fall of the Orgueil meteorite in 1864 enabled various scientists to confirm the previous observations once again. Clöez noted in 1864: "the remarkable similarity between the elementary composition of the terrestrial humic substances and that of the carbonaceous matter in the Orgueil meteorite".

Working on a sample of the Orgueil chondrite, in 1868 Berthelot detected the presence of hydrocarbons after the hydrogenation of a meteorite sample. This observation led the French chemist to the following conclusion: "whatever this carbonaceous substance is, it shows a new analogy between carbonaceous matter in meteorites and carbonaceous substances of organic origin on the surface of the Earth".

People interested in other information concerning the historical background of the search for organic matter in meteoritic samples will find all that they need in the remarkable book by Nagy [2], from which all the previous translated quotes are taken. It is interesting to observe with Nagy [2] (who himself discussed this problem with Urey) that during the nineteenth century, the scientific community was apparently not too perturbed to learn from Berzelius, Wöhler, Clöez, Berthelot and others that organic matter was present in extraterrestrial samples, even if at that time the biogenetic origin of this matter was explicitly or implicitly considered.

Surprisingly, during the first half of the twentieth century, the indigenous character of the organic matter was seriously doubted by some scientists. The paper by Spielmann [3] is illustrative of this kind of attitude: the presence of organic matter is described as an artefact due to the action of terrestrial water on hypothetical metal carbides, and also due to contamination. In 1953, Mueller [4] clearly confirmed the endogenous character of organic matter in the Cold Bokkeveld chondrite, and suggested an abiotic origin for this matter. This work can be considered as the starting point of a new interest in this field, which has evolved very rapidly since 1969, "the year" for people involved in abiotic or prebiotic organic chemistry. In fact, 1969 corresponds to the fall of two carbonaceous chondrites: Murchison in Australia and Allende in Mexico. These two meteorites have been extensively studied using modern separation and analytical techniques. The results so obtained have fully confirmed and also largely extended the conclusions reached during the previous century.

3 What Are Carbonaceous Chondrites?

The mass of the earth increases daily due to falls of extraterrestrial material. The estimates by different authors of the flux of extraterrestrial matter vary by many orders of magnitude, depending on the estimation method used. In his recent book, Dodd [5] gives an estimation of 10^2-10^3 tons per day, the largest part being in the shape of dust particles (or micrometeorites).

Most of the small amount of material (meteorites) recovered belongs to the stony meteorite class, and in this class, the chondrite subclass is the most populated. Surprisingly, iron meteorites, which are generally considered as the stereotype of meteoritic material, account for less than 10% of meteorite falls [4,5].

3.1 Classification of Carbonaceous Chondrites

Chondrites can be described as conglomerate rocks characterised by a overall chemical composition similar to the composition of the sun (with a depletion of hydrogen, helium and some other highly volatile elements). This latter characteristic is an easy way to distinguish clearly between stony meteorites and terrestrial stones.

In all but one type of chondrite, small millimeter-sized spherical silicate grains are trapped in the mineral matrix. These small grains consist mainly of amorphous silicates and are called chondrules. This latter term gives its name to the meteorite class containing chondrules, i.e. the chondrite class. The chondrites themselves are divided into three subclasses: enstatite, ordinary chondrites and carbonaceous chondrites. Carbonaceous chondrites themselves are classified into four groups: I (for Ivuna), M (for Mighei), O (for Ornans) and V (for Vigarano).

Table 1. Mean composition [wt. %] of carbonaceous chondrites

Class:	Carbonaceous (C)			
Group:	I	M	O	V
Si	10.40	12.96	15.75	15.46
Ti	0.04	0.06	0.10	0.09
Al	0.84	1.17	1.41	1.44
Cr	0.23	0.29	0.36	0.35
Fe	18.67	21.56	25.82	24.28
Mn	0.17	0.16	0.16	0.16
Mg	9.60	11.72	14.52	14.13
Ca	1.01	1.32	1.57	1.57
Na	0.55	0.42	0.46	0.38
K	0.05	0.06	0.10	0.03
P	0.14	0.13	0.11	0.13
Ni	1.03	1.25	1.41	1.33
Co	0.05	0.06	0.08	0.08
S	5.92	3.38	2.01	2.14
H	2.08	1.42	0.09	0.38
C	3.61	2.30	0.31	1.08
Fe^0/Fe_{tot}	0.00	0.00	0.09	0.11
No. of Samples	3	10	5	7

Oxygen, not reported in the table, makes up the difference from 100%.
From Ref. [5], p. 19

Table 1 gives the chemical composition of the four groups of carbonaceous chondrites, as published by Dodd in 1981 [5]. From Table 1, it appears that the C-content of the carbonaceous chondrite in group I (CI) is larger than 3% (while it is negligible in ordinary chondrites). Nevertheless, the C-content is not by itself the best means of identification of carbonaceous chondrites: some enstatites contain 0.84% of C. A better chemical descriptor of carbonaceous chondrites is the Fe^0/Fe_{tot} ratio, which shows the highly oxidized state of the mineral matrix. Fe^{3+} is found in some carbonaceous chondrites in the form of magnetite (Fe_3O_4). Other ratios, like Mg/Si, Al(Ti, Ca)/Si and Fe/Si, show subtle differences between classes and even between groups [5].

The classification of chondrites on the basis of their elemental composition is recent with respect to the more "traditional" classification based on textural and mineralogical differences [6]. In the case of carbonaceous chondrites this classification was revised by Wasson in 1974 [7]. The petrological type (from 1 to 6 even if some authors also use 7) is intended to indicate the degree of equilibration and metamorphical recrystallisation. So 1 indicates the least-equilibrated and 6 the most-equilibrated chondrites. Following this classification, CI carbonaceous chondrites are the only members of petrological type 1 and until recently, were in fact considered as the least metamorphised chondrites (and therefore as the least metamorphised meteorites). It is interesting to observe that type 1 chondrites do not contain chondrules: the mineral matrix is fine-grained and black and the bulk water content is of the order of 20%. CM carbonaceous chondrites (type 2) contain very sharply defined chondrules and a lower water content (2%–16%). Until recently, they were considered as slightly more metamorphized than CI 1 chondrites. Following this kind of classification, CV 3 chondrites were described as more metamorphized (water content 0.3%–3%). In 1979 McSween[8] pointed out that type 3 chondrites might in fact be better preserved (in the sense of less equilibrated) than type 1 or type 2 chondrites, and this point of view is explicitly accepted by Wasson in his recent book[9]. Following this description, silicate minerals like olivine (which are present in CV 3 chondrites) have been altered by hydrothermal processes into CI 1 or CM 2 chondrites and converted into clay-like minerals. For people interested in the presence and the origin of organic matter in carbonaceous chondrites, the CI 1, CM 2 and CV 3 groups are the most interesting. The Orgueil and Alais meteorites are members of the CI 1 group, and the Murchison and Allende meteorites are examples of CM 2 and CV 3, respectively.

At this stage it is important to give some indication of the composition of the mineral matrix. (For full details, see Refs. [2] and [5].)

3.2 CI, CM and CV Carbonaceous Chondrites

CI chondrites are characterised by a fine-grained (10–100 µm) hydrous silicate incorporating magnetite. This hydrous silicate is 30% similar to terrestrial montmorillonite but the dominant phyllosilicate is a septechlorite, which itself is structurally like serpentine $[Y_6(Z_4O_{10})(OH)_8$, where $Y = Fe^{2+}, Mg^{2+}; Z = Si, Al, Fe^{3+}]$. Some metallic sulphides are present in very low concentrations. High-temperature minerals (silicates) are present as fragments in the form of olivines and pyroxenes. CI chondrites show veins and vein minerals which are considered to be proof of alteration by water[8,10].

CM chondrites constitute an abundant group (15 members), and numerous studies on one member of this group (the Murchison meteorite) have produced a good knowledge both of the mineral matrix of this kind of chondrite and of the crystal chondrules and aggregates which are immersed in the matrix. Again, as in the case of CI, hydrous silicates predominate as constituents of the matrix in the form of montmorillonite and Al-poor septechlorite. Veins seem to contain sulphates, and gypsum has been characterised in Murchison. A magnetic mineral is also present in various CM chondrites but has not yet been fully characterised[5]. In the case of CM chondrites, the total content of high-temperature minerals is much higher than in the case of CI chondrites (approx. 50 wt. %). They are present as chondrules, crystals and crystal fragments, melted agregates and xenoliths (exotic chondritic material). By way of a brief description of these minerals, it seems sufficient to say that silicates predominate (olivines, pyroxenes). As in the case of chondrules found in ordinary chondrites, the chondrules in CM look like frozen melt droplets.

CV chondrites have also been extensively studied. The Allende meteorite, which fell in February 1969, is an object of such exceptional interest that the number of papers concerning this particular CV 3 chondrite are more numerous than all those published on other CV meteorites. The matrix in CV chondrites (Al Rais and Renazzo excepted) is essentially composed of fine-grained iron-rich olivine but, in Allende, sulphides are abundant (a few percent). Moreover, some CV chondrites contain fine-grained xenoliths of unknown origin as part of the matrix. Chondrules are abundant (44 vol. %) and consist mainly of olivine with grains of metal, sulfides and magnetite. Aggregates of olivine are also detectable, but aggregates (also called inclusions) containing calcium and aluminium silicates are of greater importance (abbreviated CAI). Some of them, especially in the Allende meteorite, contain elements with anomalous isotopic compositions. The study of the so-called white inclusions in the Allende meteorite was at the centre of a very interesting scientific debate: are these CAI of presolar origin and have they been injected as such into the protosolar nebula [6,11,12]? The explanation based on an initial heterogeneity of the protosolar nebula, first suggested by Clayton et al. in 1973 [11], now seems to be accepted by the scientific community.

People interested in CO carbonaceous chondrites will find information in Ref. 5.

At the end of this section dealing with the question "What are carbonaceous chondrites?" and before trying to answer the question "Where do they come from and what is their age?", it seems necessary to emphasise the fact that the carbon present in the CI and CM and to a lesser extent in the CO and CV chondrites exists as elemental carbon (CI, CM, CO, CV), organic molecules (CI and CM) and an ill-defined macromolecular compound (CI, CM, CO, CV). The exact composition of this complex mixture will be discussed below. The only point that we would like to stress in this section is that organic matter is always found intimately mixed with mineral matrices.

4 Carbonaceous Chondrites: Where Do They Come From and What Is Their Age?

These questions are of prime interest. As we said in the introduction, the study of organic matter in carbonaceous chondrites is interesting because it may give some insight into the origin of our solar system. It is out of the question to discuss here in detail the train of events that is described as the formation of the solar system. Nevertheless, it is important to remember that the age of the solar system is about 4.5 Gyr (4.5×10^9 yr).

4.1 Formation of the Solar System

The sun, the planets and satellites like the moon were formed 4.5 Gyr ago as a consequence of gravitational instability in a part of a dense interstellar cloud. This particular dense insterstellar cloud no longer exists for obvious reasons, but other dense interstellar clouds can still be observed in our galaxy. Some of these dense clouds are described as star nurseries because very young stars are detectable in them. It is therefore not too difficult to imagine that the protosolar nebula was similar to a small region of one of these dense clouds. The further contraction of this small region led to the accretion of the solar system and probably also to the accretion of other stars, which unfortunately cannot be identified after such a long time. The direct relation between interstellar media (gases and dust particles) and stars is therefore evident. In an old galaxy like our Milky Way (10–18 Gyr), 90 % of the total mass is in the form of massive stars (the masses of the planets, even if they are present around a large number of stars, is negligibly small with respect to the stars' masses). Of the total mass, 10 % is in the form of gases and dust particles that form immense clouds of different types. These gases and dust particles were ejected by stars of previous generations, not only during their lives (stellar winds) but especially at the end of them. These ejecta were the building materials of second-generation stars. Matter is therefore recycled continuously but, during this recycling process, matter changes. At the beginning, first-generation stars contain only hydrogen, helium and some other very light elements. Second-generation stars (like our sun, which might be a third-generation star) contain heavier elements formed by fusion, neutron capture or spallation in the first-generation stars.

Gravitational instability can occur in a cloud characterised by high density and low temperature (Jean's criterion). More precisely, the mass of the cloud must be of the order of 20 $M_\odot$ (where $M_\odot$ is the solar mass) if the density (expressed as the number of dihydrogen molecules per cubic centimeter) is around 10^3 and the temperature around 10 K. Such a density is precisely what is observed in so-called dense clouds. These dense clouds are also regions of space where a lot of complex molecules are detected by spectroscopy, and where dust particles are observed. Readers interested in interstellar chemistry will find an excellent review of the subject in the recent book by Duley and Williams [13].

What is important in the context of chondrite chemistry is the fact that the probable presence in the protosolar nebula of dust particles and complex organic molecules is evidenced. This does not mean that all the organic matter detected in carbonaceous chondrites is necessarily molecules still present in the protosolar nebula. Readers interested in details of the formation of the solar system and the accretion phenomena will find a lot of information in the papers by Larimer [14] and Cameron [15].

4.2 The Protosolar Nebula: A Homogeneous or Heterogeneous Medium?

An important point in this context is that before the gravitational collapse, the temperature of the cloud was very low (below 50 K) and the cloud contained many complex molecules and dust particles. During the process of collapse, including the formation of the protosolar nebula, the temperature increased. Until recently, it was considered that the increase of temperature was so great that it first vapourised all the molecules forming the dust particles and even dissociated all the complex molecules. Temperatures as high as a few thousand kelvin were considered in some accretion models (all matter is vapourised at 2000 K and at a pressure in the range of 10^{-6}–10^{-2} atm) [14]. In such models, the protosolar nebula achieved a state of "chemical homogeneity". Nevertheless, as pointed out by Clayton et al. [11] (and many authors since) the observation of isotopic anomalies in chondritic samples seems to indicate that a state of complete homogeneity never existed. The injection of matter from other star systems into the protosolar system is a very likely process which might account for some of the heterogeneities [16]. Moreover, a temperature gradient between the centre of the accretion disc (where the sun was forming) and the external part of the disc was certainly present at the beginning of the solar system. This temperature gradient acting together with gravitational effects might be at the origin of matter discrimination along the disc radius. The consequence of such discrimination can easily be observed if we compare the average density of the internal and the external planets. This mechanism is obviously inefficient as far as explaining the remaining isotopic anomalies is concerned, but is an observation of prime interest, from the chemist's point of view. It proves clearly that turbulences due to the high velocities of collapsing matter did not constitute a perfectly efficient process of homogenisation.

The heterogeneities of the protosolar nebula are considered as very significant by some authors [17], and as quantitatively not important by others [9]. Nevertheless, it remains a fact that the model of a homogeneous accretion disc where equilibrium reaction takes place has now been replaced by a much more subtle description where a lack of homogeneity and equilibrium are facts that must be taken into account. This problem will be discussed below.

4.3 Carbonaceous Chondrites: Witnesses of the Accretion Process

Chondrites, like all meteorites, are fragments of larger bodies. Nevertheless, in the case of carbonaceous chondrites, the size of the parent body (or bodies) was such that the accretion temperature always remained sufficiently low to avoid any differentiation of matter on the basis of its density. The accretion temperature of the parent bodies of carbonaceous chondrites can be estimated using various cosmothermometers (based on careful measurements of isotopic ratios). All the values are consistent: the temperature is in the range of 350–400 K for CI, and slightly higher for CM and ordinary chondrites [1]. Higher accretion temperatures and differentiation are well documented for the terrestrial planets, which are characterised by a dense core and less-dense surrounding envelopes. Incidently, this kind of differentiation implies a melting or, at least, a partial fluidity of the body (e.g. the terrestrial planets, the moon). This melting, together with the effect of universal attraction, is at the origin of the spherical shape of these objects. Many small asteroids (a few kilometers across) are not spherical at all (for example Phobos and Deimos, the two presumably captured asteroids, which are presently satellites of Mars). The parent bodies of carbonaceous chondrites (and more generally chondrites) are probably undifferentiated (and non-spherical) asteroids.

Chondritic matter can therefore be described as frozen protosolar matter slightly affected by differentiation processes (which are obviously accompanied by very large metamorphism). Chondrites are valuable remnants of the state of matter prevailing 4.5 Gyr ago! This is the major reason why their study is of such great interest. It would be an error to consider that all molecules now detected in chondrites have remained unchanged for 4.5 Gyr. Some of them might be unchanged and many of them might have been slightly modified. As we have seen above, the CI and CM parent bodies have suffered some partial hydrothermal metamorphism during their lives, which has probably transformed some silicates into phyllosilicates. Nevertheless, these transformations were essentially isochemical. The average composition of CI and even CM chondrites remains very similar to the solar composition, which, due to the solar mass, is considered to be very similar to the composition of the protosolar nebula.

4.4 Age of Carbonaceous Chondrites and Their Parent Bodies

The dating of chondrites gives a radiometric age in the range of 4.45–4.53 Gyr [6] (based on the $^{87}Rb/^{76}Sr$ system). For Wasson [9], the higher value obtained on the less-metamorphized chondritic samples gives the best estimate of the age of the solar system (4.53 $\pm$ 0.02 Gyr). Following this author, many differentiated meteorites (fragments of larger parent bodies in which differentiation has occurred) are of approximately the same age. This proves that the heating and melting processes of the parent bodies took place during the first 0.1 Gyr of the existence of the solar system. Some chondrites are much younger: the parent body (bodies?) was (were) involved in a heating process (all the nuclear clocks were reset to zero) in the relatively recent past. A possible cause could be a collision between the parent body and another asteroid.

This 4.53 Gyr represents the time that has elapsed since the last high-temperature event, which determined the time t = 0. For the oldest meteorites, the formation interval can also be estimated. Short half-life (1–100 Myr) radionuclides were present in the young parent body: the daughters can be detected. This is the case of ^{129}Xe (daughter of ^{129}I with a half-life of 16 Myr). Since ^{129}I was frozen in the parent body matrix, this means that the accretion process took a short time (a few million years).

The cosmic ages of meteorites can be determined: these correspond to the time elapsing between the break up of the parent body and their fall to earth. During this period (a few million years), a small meteorite (generally a meter or less) is irradiated by galactic cosmic rays, and traces of this irradiation can be detected. This observation leads to an estimate of its cosmic age [9]. Finally, terrestrial ages can also be measured, but we will not consider this problem here: the interesting carbonaceous chondrites like Orgueil, Alais, Murchison and Allende were collected immediately after their fall, so their terrestrial age is negligibly small [2].

At the end of this section, two final points merit some attention. The first one concerns the location of the parent bodies of the chondrites. The asteroid belt between Mars and Jupiter is one possibility favoured by many authors, but some other possibilities exist (such as the families of asteroids crossing the earth's orbit). Asteroids themselves are of many different types and some of them are probably extinct comets (after too many passages at the perihelion). Some chondrites could be fragments of these extinct comets, but this hypothesis is not the most probable one. People interested in this problem will find information in Refs. 2, 5 and 9.

The final point that generally leads to questions from non-specialists concerns the heating of a meteorite when it enters the atmosphere. This problem is discussed in detail by Nagy [2]. Due to friction with the upper atmosphere, ablation processes take place. The loss of mass is great and the heating of the object leads to the fireball phenomenon. A long dust tail is observed. Generally, the meteorite breaks up into many fragments. When these fragments are collected, they show a very thin fusion crust. Due to the refractory character of the silicate matrix, the heating process only affects the surface of the meteorites. All important analyses (chemical, mineralogical) are performed on samples of meteorites interiors which never undergo any appreciable heating during their journey through the atmosphere. Analyses, and in particular analyses of organic matter, are therefore performed on genuine samples, uncontaminated and not overheated during their fall.

5 Organic Matter in Carbonaceous Chondrites with Special Emphasis on the Murchison (CM2) Meteorite

Although organic matter has been detected and analysed in many chondritic samples, we will focus our attention primarily on the Murchison meteorite.

The fall of a carbonaceous chondrite in September 1969 near Murchison (85 miles north of Melbourne, Australia) was an event of great importance for scientists involved in meteoritic research. In 1969, many laboratories were well equipped to analyse the lunar samples, and interest in extraterrestrial matter was at its height when suddenly 83 kg of a carbonaceous chondrite were available. The biggest fragment

Table 2. Carbonaceous compounds in the Murchison meteorite

Compounds	Abundances (whole rock)		Ref.
	[%]	[ppm (µg/g)]	
Acid insoluble phases (polymer type)	1.45		[20]
Carbonates/CO_2	0.1–0.2		[20]
Hydrocarbons: aliphatic		12–35	[19]
aromatic		15–28	[21]
Acids: monocarboxylic (C_2—C_8)		~330	[22,23]
dicarboxylic (C_2—C_9)		n.m.	[24]
α-hydroxycarboxylic (C_2—C_5)		14.6	[25]
Amino acids		10–22	[19,26,27]
Alcohols: primary (C_1—C_4)		11	[28]
Aldehydes (C_2—C_4)		11	[28]
Ketones (C_3—C_5)		16	[28]
Amines: primary (C_1—C_4)		10.7	[29]
secondary (C_2—C_4)		n.m.	[29]
Urea		25	[30]
Basic N-heterocycles (pyridines, quinolines)		0.05–0.5	[31]
Pyrimidines		0.06	[32]
Purines		1.2	[33]
Carbynes		n.m.	[34]

n.m.: not mentioned

Table 3. Amounts of low molecular weight compounds found in extracts of the Murchison meteorite. Data are expressed in nanomoles per gram. (the list is not exhaustive)

R—	R—H	R—OH	R—NH$_2$	R—NH—R	R—CHO	R—CO—R' R'=CH$_3$	R—CO—R' R'=CH$_3$—CH$_2$—	R—COOH
H—					+			
CH$_3$—	8.9	~156	~194	+	~159	~103		1700
CH$_3$—CH$_2$—	8.5	~65	~44	+	~517	~28	~35	1830
CH$_3$—CH$_2$—CH$_2$—	8.9	~33	~15		~14	~58		380
CH$_3$—CH— ⎸ CH$_3$			~22	+				500
CH$_3$—CH$_2$—CH$_2$—CH$_2$—	5.5	~14	~3					120
CH$_3$—CH—CH$_2$— ⎸ CH$_3$	4.3	+						100
CH$_3$—CH$_2$—CH— ⎸ CH$_3$			+					120
CH$_3$—C— ⎸ (CH$_3$)(CH$_3$)			+					(+)
CH$_3$—CH$_2$—CH$_2$—CH$_2$—CH$_2$—	~12							60
CH$_3$—CH—CH$_2$—CH$_2$— ⎸ CH$_3$								70
CH$_3$—CH$_2$—CH—CH$_2$— ⎸ CH$_3$								(+)
CH$_3$—CH$_2$—CH$_2$—CH— ⎸ CH$_3$								(+)
CH$_3$—CH$_2$—C— ⎸ (CH$_3$)(CH$_3$)								(+)
CH$_3$—CH—CH— ⎸ ⎸ CH$_3$ CH$_3$								(+)
(CH$_3$—CH$_2$)(CH$_3$—CH$_2$)CH—								(+)
CH$_3$—C—CH$_2$— ⎸ (CH$_3$)(CH$_3$)								
Ref.:	23)	28)	29)	29)	28)	28)	28)	22, 23, 38)

+: positively identified; (+): tentatively identified
The numerical values taken are the highest given in the literature.

Table 3. (continued)

—R—	$R\!<^{COOH}_{COOH}$	$\alpha R\!<^{OH}_{COOH}$	NH₂—R—COOH*				
			α	β	γ	δ	ε
—	+						
$-CH_2-$	+	21	81				
$CH_3-\overset{\mid}{C}H-$	+	65.5	39.3				
$-CH_2-CH_2-$	+			4.5			
$CH_3-CH_2-\overset{\mid}{C}H-$		22	10.6				
$CH_3-\overset{\mid}{\underset{CH_3}{C}}-$		21.1	24.2				
$CH_3-\overset{\mid}{C}H-CH_2-$	+			6.8	+		
$-CH_2-CH_2-CH_2-$	+				+		
$CH_3-CH_2-CH_2-\overset{\mid}{C}H-$		13.5	+				
$CH_3-\overset{\mid}{C}H-\underset{CH_3}{C}H-$		4.2	17				
$CH_3-CH_2-\overset{\mid}{\underset{CH_3}{C}}-$		4.2	+				
$CH_3-CH_2-\overset{\mid}{C}H-CH_2-$ (CH₃)				≦1; ≦1			
$CH_3-\overset{\mid}{C}H-\underset{CH_3}{C}H-$	+		≦1				
$-CH_2-\overset{CH_3}{\underset{CH_3}{C}}-$	+			~2; ≦1			
$-CH_2-CH_2-\underset{CH_3}{C}H-$					~5; ~5		
$-CH_2-\underset{CH_3}{C}H-CH_2-$	+				~5		
$-CH_2-CH_2-CH_2-CH_2-$	+						~5
$\overset{CH_3}{\underset{CH_3}{>}}CH-\overset{CH_3}{\underset{\mid}{C}}-$			8				
$CH_3-CH_2-CH_2-\overset{\mid}{\underset{CH_3}{C}}-$			6				
$\overset{CH_3-CH_2}{\underset{CH_3-CH_2}{>}}C<$			6				
$CH_3-\overset{CH_3}{\underset{CH_3}{C}}-CH<$			≦4				
$CH_3-CH_2-\overset{\mid}{\underset{CH_3}{C}}H-CH<$			≦4				
$CH_3-CH_2-CH_2-CH_2-CH<$			≦2				
$CH_3-\overset{\mid}{\underset{CH_3}{C}}H-CH_2-CH<$			≦4				
Ref.	24)	25)	19, 26, 27, 39–43)				

* Hydrolyzed water extract.

95

weighed 7 kg [18], while many others were in the range of a few grams. Many fragments were gathered immediately after the fall, which was observed by many eyewitness. The risk of contamination of the interior of the samples was therefore strongly reduced. All these conditions led to the first announcement in the literature in December 1970 of the presence of aminoacids and indigenous hydrocarbons in an extraterrestrial sample [19].

The indigenous nature of the aminoacids was unambiguously demonstrated by the fact that they were in the racemic form. Moreover, some of them were unknown as constituents of proteins. Since 1970, many other analyses have substantiated these first results and many other organic molecules have been identified. What we beleive to be an exhaustive list is given in Table 2.

5.1 Macromolecular Substances

Before commenting on some interesting features that appear in Table 2, it is important to note that the organic matter is not only present as volatile or extractable individual molecules. As they appear in Table 2, 70%–95% of the organic matter in these carbonaceous chondrites is in the form of an ill-defined macromolecular substance (frequently called polymer). This polymer is intimately imbricated in the mineral matrix. The analysis gives the following values for elemental composition: C 73.1%, H 4.27%, N 2.68%, ash 3.97%, O + X 15.94% (X = Cl, F) [35]. The polymer seems to be highly unsaturated with aromatic and heteroaromatic ring systems bridged by short aliphatic chains [35] and ether linkages [36].

Some of the aminoacids and, to a greater extent, some of the nitrogen heterocycles detected as free molecules in the chondritic matter might be fragments of the polymer. This fragmentation may have been caused by hydrolysis during extraction procedures [35].

5.2 Hydrocarbons

The relative amount of aromatics, alkanes and alkenes seems to depend strongly on the molecular weight: for lower weights (below C_{11}) aromatics seem to dominate, while the reverse seems true for higher molecular weights [1]. In a series of analyses performed in 1972 [37] on the volatile fraction released at 150 °C, alkanes, branched alkanes, benzene and toluene were easily detected but n-alkane seemed to be absent. Nevertheless, using another extraction procedure, Yuen et al. [23] were able to isolate the low molecular weight hydrocarbons (see Table 3), and they found that saturated hydrocarbons were more abundant than alkenes; moreover benzene was three times more abundant than the CH_4 recovered (in mol. g^{-1}). Such a result proves unambiguously that the organic matter containing only C and H is far from equilibrium (taking into account the accretion temperature) [23]. This problem will be discussed below.

In the heavy alkane fraction, normal alkanes predominate over methyl- and di-methylalkanes. Moreover, the number of detected alkanes was negligible in relation to the number of possible isomers. Six isomers of the C_{16} saturated hydrocarbon were observed while the number of possible isomers was around 10^4 [37].

The same comment can be made with respect to aromatic hydrocarbons. The relatively large amount of aromatic hydrocarbons is associated with a limited variety of structures [1,21,37,39] from benzene to coronene.

5.3 Carboxylic Acids

The study of monocarboxylic acids in Murchison is described in three papers [22,23,38]. As reported in Tables 3 and 4, all possible acids containing 2, 3, 4 and 5 carbon atoms were detected. Moreover 7 of the 8 possible isomers of the C_6 acid were also observed.

The acids with an even or an odd number of C-atoms were present in a similar amount. The same remark is valid for linear and branched isomers. This distribution pattern rules out terrestrial contamination [38].

Monocarboxylic acids might be present in Murchison as free acids, esters or salts: nothing is known about this aspect of the problem [38]. They are relatively abundant in Murchison, more abundant than aminoacids and hydrocarbons. The relative concentration of propanoic acid with respect to propane is 205, while the ratio is 22 if propanoic acid is compared to glycine. So far, 17 dicarboxylic acids have been identified in Murchison. The lack of reference compounds precludes the identification of many others [24].

The relative concentration of dicarboxylic acids with respect to aminoacids is higher (by one or two orders of magnitude). As in the case of monocarboxylic acids, every possible isomer seemed to be present, ranging from C_2 to C_5 molecules. In the case of chiral molecules, the two enantiomers coexisted in nearly equal concentration. Oxalic acid was detected as calcium salt, but the state of the other dicarboxylic acids in the Murchison meteorite remains an open problem [24]. Dicarboxylic acids, as monocarboxylic acids, seem to be the result of synthetic pathways that give mixtures at random [44,45].

The Murchison meteorite contains α-hydroxycarboxylic acids [25,46]. Seven of them

Table 4. Isomeric distribution of acids in Murchison

Number of C atoms	Acyclic primary monoamino alkanoic acids					Mono-carboxylic acids	Di-carboxylic acids	α-Hydroxy-carboxylic acids
	α	β	γ	δ	ε			
2	1(1)	—	—	—	—	1(1)	1(1)	1(1)
3	1(1)	1(1)	—	—	—	1(1)	1(1)	1(1)
4	2(2)	2(2)	1(1)	—	—	2(2)	2(2)	2(2)
5	3(3)	5(5)	3(3)	1(1)	—	4(4)	2(2)	3(3)
6	7(7)	10(0)	9(0)	4(0)	1(0)	8(7)	6(4)	7(0)
7						18(2)	14(4)	
Ref.	19,26,27,39–43)					22,23,38)	24)	25)

The number of isomers identified in Murchison extracts is given in brackets.

were detected. Their concentrations are similar to those observed for the corresponding α-aminoacids, with the same general abundance pattern (see Table 3).

The presence of α-hydroxycarboxylic acids together with α-aminoacids could lead to an estimate of the local concentration of ammonia when these molecules were synthesised. Such an estimation method implies the assumption that the syntheses of the two classes of molecules were simultaneous and started from the same organic substrate, i.e. aldehydes [25]. From aldehydes, aminoacids can be obtained by the Strecker synthesis (aldehyde, HCN, NH_3 in aqueous solution), while hydroxyacids can be synthesised from the cyanhydrin synthesis (aldehyde + HCN) followed by a hydrolysis. Nevertheless, it must be emphasised that all aminoacids detected in carbonaceous chondrites cannot be obtained by the Strecker synthesis. This remark limits the interest of the previous arguments concerning the concentration of NH_3 during the accretion phase.

5.4 Aminoacids

The aminoacids are certainly the molecules that excited people's imaginations most, whether they were scientists or otherwise, when they were discovered in chondrites. A priori, the presence of aminoacids is not so surprising if we know that these molecules can easily be made by prebiotic simulation experiments. It seems easy to obtain aminoacids by combining together H, C, N and O in various molecular forms. Until now, 34 aminoacids have been fully identified in the Murchison meteorite [19,26,27,39–43] (see also Table 3). Only 8 of them are commonly found in proteins. When chiral, they are obtained as racemic mixtures (with perhaps one exception). The protein aminoacids not detected are those containing OH, SH (and S—S) groups, aromatic or hetero-aromatic moieties. The exchange of some H atoms for D after treatment by D_2O gives rise to an interesting suggestion concerning the existence of an aminoacid precursor [47] but it appears that the deuterium incorporation can be explained by a simple exchange process [27]. This exchange is accompagnied by partial racemisation, as is normal for a process that implies the reversible cleavage of the α C—H bond. The possible efficiency of this reversible cleavage as a racemisation process could be considered as an explanation for the existence of chiral aminoacids as enantiomeric mixtures. In this context, isovaline (without an α C—H bond) is of particular interest. This molecule cannot racemise by the mechanism just described. However, it is present as a racemic mixture [R(—) 52.2%, S(+) 47.8%] [48]. This observation is compatible with an abiotic synthetic pathway without any enantioselectivity. Nevertheless, Bonner et al. [49,50] were able to show that isovaline itself racemises when submitted to radiolysis. This observation has been discussed in the literature [43]. The authors concluded that it was difficult to prove that the radiolytic cleavage and the consequent isomerisation took place in the chondrite parent body or, after fragmentation, in the chondrite itself when submitted to cosmic rays. Too many unknowns exist concerning the real doses experienced by the organic matter. It remains true that the isomeric distribution of the aminoacids and the relative quantity of the carboxylic acids with respect to the aminoacids is not incompatible with radical reactions. The loss of the amino group leading to carboxylic acids is a conceivable degradation pathway not incompatible with the relative amount of amino and carboxylic acids. Moreover the presence of α-, β-, γ- and δ-aminoacids and the corresponding hydroxyacids suggests radical reactions, whether or not they are initiated by radiolysis.

To come back to the chirality problem, Engel and Nagy [51], claimed in 1982 that several of the protein aminoacids were non-racemic (with an excess of L isomer) while the non-protein aminoacids were racemic. This result is certainly interesting but it is very unfortunate that the enantiomeric excess was not observed for the D isomer! The excess of the L isomer is obviously what could be predicted in cases of contamination. Moreover, even an excess of D could be explained by contamination due to the presence of D α-aminoacids in some microorganism walls. On the basis of a careful analysis of the results of Engel and Nagy, Bada et al. [52] concluded that these results were most likely explained by terrestrial contamination. This last paper was criticised by Engel and Nagy in a short reply [53]. It seems useful to reproduce here the last sentence of their note: "We propose that it would be more productive to conduct additional experiments on meteorites aminoacids and weigh the validity of the resultant interpretation against each other to obtain a higher level of comprehension."

Before leaving the aminoacids problem, it is interesting to note that aminoacids have been detected in carbonaceous chondrites found in Antarctica. The risk of contamination is much less important in Antarctica than in Australia and this is one of the reasons why these studies were undertaken. They fully confirm the results obtained on Murchison [54,55], even if in one CM carbonaceous chondrite the amino acid content was only 10 % of what was observed in Murchison [56,57]. The contamination is in fact lower than in Murchison; the aminoacid content was very similar for samples taken near the surface of the Antarctica chondrites or from their bulk. On the other hand, all the significant analyses on Murchison were performed on samples from the interior of the meteoritic fragments due to the high degree of surface contamination. In the case of the Allende meteorite, which has the same terrestrial age as Murchison, contamination was found to extend to a depth of more than 5 mm below the surface [52].

5.5 Heterocycles

Nitrogen-containing heterocycles are of obvious interest in the context of prebiotic chemistry. This is the reason why we will now consider this class of derivatives. The first so-called evidence for the presence of 4-hydroxypyrimidine and 4-hydroxy-methyl-pyrimidine in Murchison was published in 1971 [58]. These results were divergent from those of previous studies performed on Orgueil samples where adenine, guanine, s-triazines and guanylurea were detected [59,60].

In 1975, Hayatsu et al. [31] tried to elucidate some of the divergences between the results obtained by their own group working on an Orgueil sample, and those obtained by Folsome et al. [58] on a Murchison sample. Also, working on two Murchison samples, they were unable to detect hydroxypyrimidines by mass spectroscopy but confirmed the presence of melanine, guanine, adenine, triazine. The discrepancies between the results obtained by two high-level groups is interpreted by the authors [30] as due to the extraction procedure. A mixture of $H_2O/HCOOH$ was used by Folsome et al. [58], while HCl (3–6 M) was used by Hayatsu et al. [30]. N-Heterocycles probably occurred as fragments of the kerogen and were obtained by cleavage under drastic conditions. This last remark illustrates one important difficulty in the interpretation of the results of analyses conducted on heterogeneous material like chondritic samples. The contradictory nature of all these results led the Schwartz group in Nijmegen to start a series of very careful studies on the search for purines and pyrimidines in the Murchison meteorite [31–33, 61, 62]. The HCOOH extracts gave purines, i.e. hypoxanthine, xanthine, adenine, guanine. The pyrimidimic compound uracil was also detected in Orgueil, Murchison and Murray, while thymine seems absent. Both in the H_2O and HCOOH extracts, the concentration of uracil was of the order of 30 ng/g sample [32]. If present, hydroxypyrimidines were characterised by a very low concentration (upper limit 10 ppb).

Melanine, ammelide and cyanuric acid were not detected in the HCOOH extracts and the identification of s-triazines previously reported does not prove that s-triazines are indigenous. This kind of compound can easily be formed during the experimental and analytical procedures. Nevertheless, this last conclusion is not definitive. As pointed out by Stoks and Schwartz [33], the carbonaceous chondrites are far from homogenous. Important variations from sample to sample exist, even if all these samples are fragments of the same chondrite. Moreover, and as previously reported, the sample treatment is of prime importance. It can lead to the formation of secondary products or liberate molecules unobservable after milder treatment. To conclude, it seems clear that nitrogen heterocycles are present in carbonaceous chondrites and that pyrimidine derivatives are less abundant than purine derivatives.

5.6 Elemental Carbon

Carbon in the carbonaceous chondrites does not exist as polymer or organic molecules alone. Carbonates are also present in relatively small amounts [20, 23] and the same is true for elemental carbon. Elemental carbon seems to exist as carbynes (triple-bonded allotropes of carbon). At least three types of carbynes have been described in Murchison [34] but these results were questioned in 1982 by Smith and Buseck [63]. According to these authors, sheet silicates mixed with elemental carbon could be misidentified as carbynes in X-ray diffraction patterns. These particular carbonaceous phases (carbynes or otherwise) and other carbonaceous phases (polymer and amorphous carbon phases called C-α and C-β) are carriers of noble gases trapped in the chondritic material. Some of these "carbynes" seem to be condensates from the protosolar nebula while others are probably of presolar origin [34].

At the end of this section devoted to the results of the numerous analyses of organic matter in carbonaceous chondrites, and particularly in Murchison, it is important to emphasise the great variety of molecules found and also the many problems posed by these analyses. Due to the concentration of the organic molecules in chondritic matter, the analyses were generally performed at the microgram level. Serious risks of contamination or misinterpretation always exist. This implies a need for both maximum precautions and a maximum of attention on the part of those who perform the experiments and those who read the papers.

Since there is a great variety of molecular structures, this gives rise to major problems with respect to the origin of these molecules. Nevertheless, the question of their origin is obviously the interesting one. Many papers have been devoted to this subject and we will discuss this problem below. Before this, we would like to give some information on isotope distribution in organic matter found in carbonaceous chondrites.

6 Isotope Distributions in Organic Matter Constituting Carbonaceous Chondrites: Analytical Data

The search for the origin of organic molecules in carbonaceous chondrites involved the use of all the information that could be considered to be the signature of the synthetical processes. A superficial analysis could lead to the conclusion that the molecular structures themselves are the best signatures. It is true that the information content of complex structures such as coronene or valine is far from negligible. Without doubt, a structure is a signature of a synthetical process, but different synthetical pathways can lead to the same structure. Isotopic distributions are another type of signature which, from our point of view, is probably more significant. This is the reason why a section is devoted to this subject.

As pointed out by Pillinger [64], the first study of stable isotope abundances in chondritic samples was published in 1934. Since then, the interest in studying isotope distribution has become greater and greater. Many anomalies have been found at the level of light elements like H, C, N and O, and we will focus our attention on these elements because of their role in organic chemistry. The isotope abundance for one particular element is sometimes very different, depending on the microsample analysed or even (and much more interestingly) on the kind of molecules of which the element forms part. The so-called isotope anomalies are direct proof that the solar system is far from homogeneous in terms of isotope distribution.

One of the first aims of measuring isotope abundance was the search for proof of the extraterrestrial origin of the light elements found in meteorites [65]. The observation [66] of a $^{13}C/^{12}C$ ratio greater in Orgueil carbonates than the same ratio in any known terrestrial samples led, more than 20 years ago, to the fundamental question concerning the origin of such differences. Before giving any analytical data, it is important to remember here the δ notation widely used in geochemistry, astrophysics and cosmochemistry to express isotope abundance; δ is expressed in parts per mil, and is defined as

$$\delta = \frac{(R_{sample} - R_{standard})}{R_{standard}} \times 1000 \, ,$$

where $R = \dfrac{\text{concentration of the heavy isotope}}{\text{concentration of the light isotope}}$.

The accepted standard compounds are:

For hydrogen   Standard mean ocean water (SMOW)
                           $D/H = 0.0001557$

For carbon   Pee Dee belemnite (PDB)
                           $^{13}C/^{12}C = 0.0112372$

For nitrogen   Atmospheric gas
                           $^{15}N/^{14}N = 0.0036765$

For oxygen   Standard mean ocean water (SMOW)
                           $^{18}O/^{16}O = 0.0020052$

For sulphur   Canyon Diablo troilite
                           $^{34}S/^{32}S = 0.0450451$

6.1 Carbon

Let us now briefly review the distribution of carbon in the Murchison carbonaceous compounds.

It appears clearly from Fig. 1 that the mean value of $\delta^{13}C$ characteristic of total carbon has no meaning in terms of isotopic signatures. The value of δ varies from one class of molecules to another, going from positive values ($\sim +30\%$) for aminoacids to negative values for the more reduced hydrocarbons. Moreover, the $^{13}C/^{12}C$ ratio for individual hydrocarbons and monocarboxylic acids decreases with

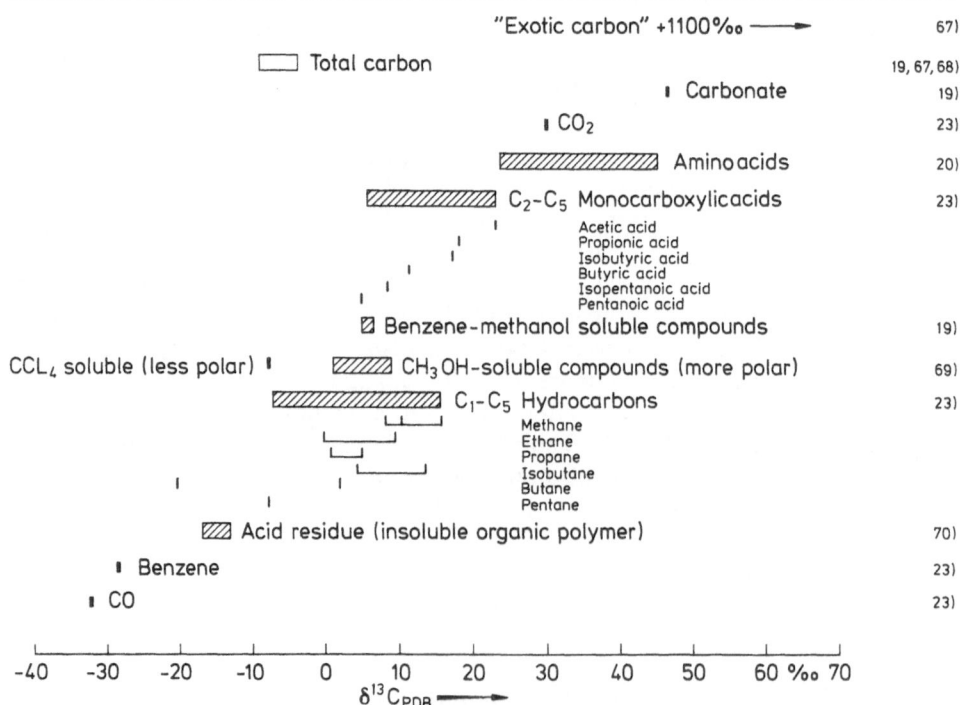

Fig. 1. Isotopic composition of carbon in Murchison (CM 2)

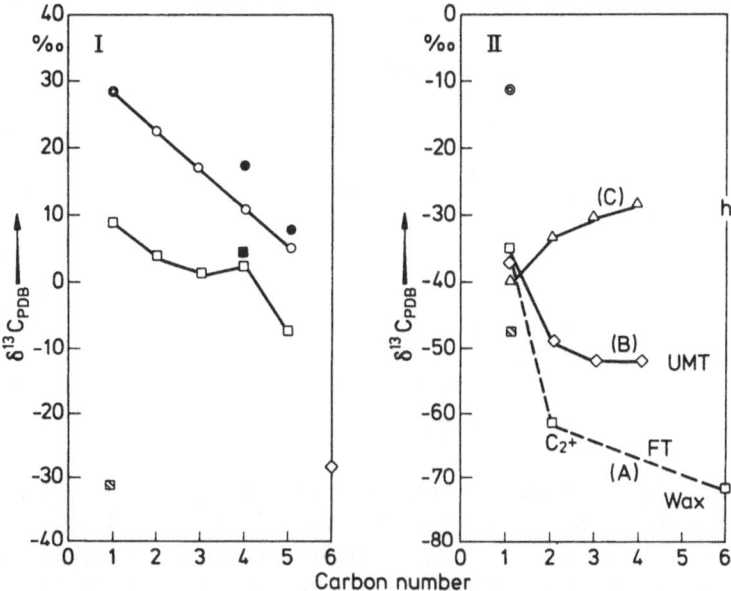

Fig. 2. I, II. Plots of the $\delta^{13}C_{PDB}$ values of light carbonaceous compounds against their carbon number. *I* Isotopic composition of individual saturated hydrocarbons ($\square$); benzene ($\diamond$); CO ($\boxminus$) and individual monocarboxylic acids ($\bigcirc$) in the Murchison meteorite, $\delta^{13}C$ values are given against carbon numbers; e.g. 1 denotes methane ($\square$) or CO_2($\circledcirc$), 2 ethane or acetic acid, etc. Branched isomers are indicated by filled symbols. (From Ref. [23].) *II* (*A*) Results of laboratory experiments where hydrocarbons ($\square$) and CO_2 ($\circledcirc$) were synthesised by a FT reaction (CO/H_2 = 1; T = 400 K; P = 760 mmHg; 320 h in the presence of a cobalt catalyst). Initial CO: $-38.6^0/_{00}$, final CO: $\boxminus$. (From Ref. [71].) (*B*) Results of laboratory experiments where hydrocarbons ($\diamond$) were synthesised by a UMT reaction starting from a methane atmosphere ($\delta^{13}C = -38.6^0/_{00}$ and P = 200 mmHg) using an electrical discharge. (From Ref. [72].) (*C*) Results of laboratory experiments: CH_4, C_2H_6, C_3H_8 and C_4H_{10} ($\triangle$) were obtained by the thermal decomposition of hexane at 773 K. (The letter h identifies the initial $\delta^{13}C$ value of hexane.) (From Ref. [72].)

increasing carbon number in a roughly parallel manner, with, in the case of the same number of carbon atoms, a higher δ for carboxylic acid than for hydrocarbons. Figure 2I gives the isotopic compositions of the saturated hydrocarbons and monocarboxylic acids plotted against carbon numbers, with CO_2 taken as the C_1-acid counterpart of CH_4 [23].

Recently, Swart et al. [67] found carbon (probably present as elemental carbon) highly enriched in C-13 in Murchison and Allende. The δ values up to + 1100 per mil correspond to $^{12}C/^{13}C$ = 42 compared to between 88 and 93 for terrestrial carbon. This exotic carbon is associated as the carrier phase with several noble gases, also characterised by anomalous isotopic distribution.

6.2 Hydrogen

If we except the earlier paper by Briggs [65], clear-cut evidence of an enrichment of deuterium relative to the hydrogen in carbonaceous chondrites was clearly demonstrated in 1980–81 [73–75]. All these authors stressed that the large values were found

Table 5. Isotopic composition of hydrogen and nitrogen in Murchison (CM2)

Compounds	$\delta D‰$	Ref.	$\delta N‰$	Ref.
Bulk	− 7 to − 65	[68]	+ 36 to + 44	[68]
Organic fraction:				
Total	+480 to +680	[70,75,76]		
CCl$_4$ soluble	~+112	[69]		[69]
CH$_3$OH soluble	~+455	[69]	+ 89	[69]
Polymer type	~+830	[70]	+ 18	[70]
Inorganic fraction:				
Total			−210 to −274	[77]
Phyllosilicates	− 90 to −135	[70,73]		
Interstellar origin?	up to +2860	[76]	<−274	[64]

exclusively for hydrogen from organic molecules ($\sim +450‰$) while hydrogen from phyllosilicates was normal with respect to terrestrial values ($-400‰$ to $+100‰$).

The δD value of $+450‰$ itself is an average value between the strongly positive values observed for the polymeric material (δD 830‰ in Murchison) and the lower values characterising the soluble organic matter (see Table 5).

Yang and Epstein [76] measured δD values up to $+2860‰$ on a sample of Murchison obtained by a stepwise pyrolysis of acid residues. These very abnormal δD values are associated with $\delta^{13}C$ which are also very high ($\sim +508‰$). These observations led the authors to entitle their paper "Relic interstellar grains in Murchison meteorite". This problem will be discussed below.

6.3 Nitrogen

The study of $\delta^{15}N$ values was also performed on CI and CM samples. In these carbonaceous chondrites, nitrogen is essentially part of organic molecules. Compared to other meteorites where nitrogen is essentially or totally present in inorganic form, CI and CM samples are enriched in ^{15}N [69].

The $\delta^{15}N$ analyses of solvent-extractable organic matter undertaken by Becker and Epstein [69] show no yields of nitrogen in the CCl$_4$ extract while $\delta^{15}N$ or the CH$_3$OH extracted organic molecules was close to $+90‰$ for Murchison.

Observations during the stepwise oxidation of the organic matter also clearly proves that the ^{15}N content is dependent on the type of host organic molecules. Acid-resistant residues from a Murchison sample submitted to oxidative treatment in order to remove the organic polymer show very low $\delta^{15}N$ values (up to $-274‰$) [64,77]. This Murchison sample contains a carbonaceous phase which is the carrier of carbonaceous chondrite fission xenon (CCF-xenon). This observation must be related to the observation by Thiemens and Clayton [78] on an Allende sample, which showed clearly that light nitrogen is associated with neon-A and CCF-xenon, and is perhaps a component of the organic carrier.

6.4 Oxygen and Sulphur

Oxygen is a very interesting element with respect to the study of isotopic effects. From the three isotopes ^{16}O, ^{17}O and ^{18}O, only ^{17}O has a nuclear magnetic moment. A priori, the isotopic distribution of these three isotopes can lead to a clear-cut differentiation between mass-dependent chemical fractionation and nuclear magnetic effects. Unfortunately, the isotope analysis of oxygen in the organic matter of carbonaceous chondrites remains essentially unknown, as is clearly pointed out in a recent review article on the subject [79]. On the other hand, a lot of space has been devoted to the isotope distribution of oxygen in the high-temperature silicate fractions [11,80,81]. The reason for this state of affairs is easily understandable if we remember that Clayton et al. [11] used the observed depletion in the heavy isotopes ^{17}O and ^{18}O in chondritic anhydrous high-temperature minerals to initiate a complete revision of theories regarding the origin of the solar system.

The last element of real interest with respect to isotope effects in organic matter is sulphur. Unfortunately, the information on this element remains very sparse [65,82].

People especially interested in the light element stable isotopes in meteorites should see the recent review article by Pillinger [64].

6.5 What is a Normal or an Abnormal Isotopic Ratio?

To end this section, we can conclude that for the major constituents of the organic matter, i.e. C, H, N, isotope distribution depends heavily on the nature of the sample analysed, on its physico-chemical treatment, on its localisation, and on its molecular content. Moreover, isotope abundances are dependent on the nature of the organic molecules analysed. In some samples, one or, more generally, several isotope distributions are so different from the average values that the elements are called "exotic". This term implies that physico-chemical effects, decay of non-extinct radionuclides and spallation reactions cannot explain these values. It therefore implies an initial isotope distribution that can be described as abnormal. As we said above, this means that the state of the protosolar nebula was never such as to lead to complete homogenisation, with the same isotope distributions in all its parts. It also means that nuclides having different origins were never completely mixed in the same locus of the protosolar nebula. By "different origins" we mean different nucleo-synthetical pathways, and therefore different "mother stars".

As we said in Sect. 4, these now generally accepted ideas were recently introduced to the literature, and are of great importance in the context of this review article. In our opinion, it is impossible to discuss the problem of the origin (or better, the origins) of organic matter in chondrites without taking into account the heterogeneity of the protosolar nebula and its consequences. In the next section, we will illustrate this last remark.

What is the criterion for labelling an element "exotic"? This question could be formulated in another way. How can we be sure that a particular isotope distribution cannot be explained by normal physico-chemical isotope effects, so making it necessary to opt for an abnormal initial isotope distribution? It must be clear that as far as we accept the model of a heterogeneous protosolar nebula, we have an "ad hoc", explanation of all kinds of isotope distributions and of all kinds of variations of this distribution from one sample to another, or even from one class of molecules to another. The "heterogeneous model" is very efficient, too efficient in fact. Of course, this is not a reason to reject it. Many papers have been devoted to mass isotope effects in geochemistry and cosmochemistry [83-85]. It seems pointless to discuss this problem here, since it is widely known to chemists and is based on well-founded arguments involving statistical mechanics. Mass isotope effects not only play a role in the case of chemical species in equilibrium (thermodynamical effect) but also in chemical kinetics. The theoretical background is always the same: mass isotope effects are related to zero point vibrational energy and partition function differences. These differences are between two ground states (in the case of equilibrium) or between a ground state and an activated complex (in the case of a kinetic process).

Moreover, mass isotope effects also play a role in phase change phenomena and diffusion processes. These effects increase when the relative difference between the isotope masses increases. It is therefore maximum for the deuterium–hydrogen pair and smaller in the case of the ^{13}C–^{12}C pair, for example. Nevertheless, the mass isotope effects are not the only ones that can play a role, and are not necessarily the most important ones from a quantitative point of view. We will first consider magnetic isotope effects, which have been known in chemistry since 1967 [86-88]. They are related to the coupling between the electron magnetic moments and the nuclear magnetic moment, and play a role in radical reactions when a bond breaking process leads to the formation of a triplet diradical in a cage (solvent cage, matrix, solid state) (see Fig. 3).

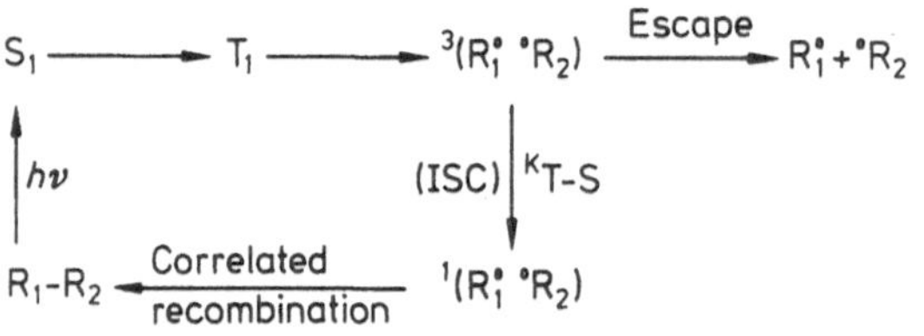

Fig. 3. Mechanism of photolysis of a caged organic compound R_1–R_2.

The recombination of the two radicals is impossible for as long as the two electronic spin moments remain parallel; during this time either one or both radicals can escape from the cage and react with other molecular species to give products different from the initial molecule. Nevertheless, if an inter-system crossing (ISC) takes place when the radicals are still in the cage, a recombination of the singlet diradicals can take place. Magnetic isotope effects are easily explained by taking into account the fact that the intercrossing rate constant is not the same if the electronic spin is localised, or partially localised, on an atom so that its nucleus may or may not have a magnetic moment. Hyperfine coupling affects the intercrossing rate constant k_{T-S}. So the isotope distribution in the recombined molecules is not identical to the distribution in the initial molecules, even if they are structurally identical. Moreover, the isotope distribution in the products of the reactions formed after the cage escape is also different from the one characterising the intial material.

The magnetic isotope effect is maximum when the escape rate is low, i.e. in highly viscous media and in solid matrices. It is therefore highly probable that this particular effect has played a role in cosmochemistry, and that some abnormalities in the isotopic composition can be explained on this basis. Some authors [89,90] have emphasised this point, and Reisse and Mullie have devoted a paper too this subject [91]. They were able to show that carbon values as high as 1000 per mil can be explained by the isotopic magnetic effect, while such values can never be obtained as a consequence of an isotopic mass effect. Previously, Thiemens and Clayton [89] had not been able to reproduce experiments performed by Galimov that proved that magnetic effects could be an efficient process in oxygen isotope fractionation. This negative result perhaps explains why magnetic effects have not been studied deeply in recent years, whereas in our view these effects could have played a key role in organic cosmochemistry. We will come back to this subject below.

Other non-mass-dependent isotope effects have been suggested recently as possible factors acting in cosmochemistry and astrophysics. The recent paper by Heidenreich III and Thiemens [92] gives a good review of the subject. Even so, these factors seem to act predominantly during the dissociation–association processes in the gas phase. They could therefore have played a role in the protosolar cloud but not during or after the accretion process. Moreover, the papers by Heidenreich III and Thiemens [92] and Navon and Wasserburg [93] are devoted to oxygen, and it is not clear whether self-shielding or symmetry effects could be efficient processes, which would explain abnormalities relative to hydrogen, carbon or nitrogen even if we limit our interest to gas-phase chemistry.

At the end of this section, we would like to emphasise some important points. The isotopic ratios observed in carbonaceous chondrites are sometimes so far from solar ratios that the simplest explanation is to postulate a heterogeneous isotopic distribution in the protosolar nebula itself. Nevertheless, the complexity of the chemical processes before and during the accretion, the variety of the chemical reactions, the diffusion processes and the other isotope differentiating mechanisms [94] that took place in the parent bodies during their long life must lead us to adopt a very cautions attitude. A great number of abnormal isotope effects is probably related to the initial heterogeneity of the isotope distribution, but non-mass-isotope effects can sometimes be so great that an effect previously called abnormal becomes "normal". The variety of isotope distribution is also proof that organic matter in carbonaceous chondrites has different origins and is not only the result of Fischer-Tropsch-type synthesis or Urey-Miller synthesis that have taken place during accretion time [1]. Whether organic or not, the matter that constitutes chondrites is an extremely complex mixture of molecules, some of which were present as such in the protosolar nebula as gaseous molecules or associated with dust particules; other were synthetised during accretion by the condensation of smaller species, while yet others were synthetised during metamorphism stages. In the next section, we will examine how organic molecules can be formed or transformed during these various stages.

7 The Origins of Organic Matter in Carbonaceous Chondrites

7.1 The Presence of Interstellar Molecules in Carbonaceous Chondrites

Much of the evidence coming from isotope anomalies in chondrite samples leads to the assumption that unchanged presolar matter is a component of carbonaceous chondrites. This idea, now broadly accepted, was introduced in 1969 by Black and Pepin [95] and was based at that time on an isotopic study of neon trapped in chondritic samples.

In 1973, Clayton et al. [11] developed the model of an inhomogeneous protosolar nebula containing solid particles, which had been preformed at a previous stage and which had never participated in a general homogenisation process in the gas phase. These solid particles were formed in a particular environment and injected into the protosolar nebula. They are therefore called presolar or, sometimes, interstellar.

As we will see, some anomalies in the isotopic composition of carbon, hydrogen and oxygen can be explained on the basis of this assumption, and we will start the discussion with the deuterium-rich matter in carbonaceous chondrites. This deuterium-rich matter is essentially present as complex macromolecules [70,73,96,97]. The carbon in these samples is essentially "normal" [76,98]. For some polymer-type fractions, the deuterium content is up to 32 times higher than the galactic value (D/H 2×10^6 in the number of atoms per cubic centimeter). High deuterium enrichments are known in interstellar molecules and the mechanism of this enrichment is fully understood. For an excellent review dealing with interstellar chemistry, see the paper by Winnewisser [99] and the previously mentioned book by Duley and Williams [13].

In 1981, Geiss and Reeves [100] suggested that the deuterium enrichment observed in the organic macromolecular compounds could be explained on the basis of a suggestion made by the same authors in 1972 [101], i.e. the survival of interstellar molecules [synthesised at low temperatures (10–50 K) by ion-molecule gas-phase reaction] in chondritic matter.

It is not easy, and probably artificial, to differentiate between a dense cloud and the protosolar nebula in its early stages, but the D-enrichment seems compatible with low-temperature–low-pressure gas phase reactions giving rise to molecules, which could have polymerised on the surface of the interstellar dust particles. Taking into account the very peculiar structure of many interstellar molecules, it has been suggested that structural determinations of chondritic macromolecular material could perhaps indicate the amount of kerogen that is of interstellar, and the amount that is of protosolar or solar origins [98]. To the extent that molecules still present in the gas phase as components of interstellar clouds have probably survived the accretion process, it becomes highly probable that unmodified (or slightly modified) dust particles form part of chondritic matter.

As we said above, grains and dust particles constitute a major component of dense interstellar clouds. Their role in the synthesis of complex molecules is accepted by many authors, but the relative proportion of gas-phase to "on grain" reactions is at the centre of an interesting debate. All authors agree on the necessary role of "on grain" reactions in the synthesis of the dihydrogen molecule, the most abundant molecule in the Universe. To return to the role of dust particles in the synthesis of complex organic molecules, Greenberg and his group [102, 103] have proposed an ingenious process that could explain the release of "on grain" synthesised molecules in the gas phase. In their model the dust particles have a silicate core and an organic mantle. This mantle could contain not only complex organic molecules formed by low-temperature photochemical reactions, but also radicals, trapped in the organic matrix. Collisions between grains could act as triggers for chain reactions. The energy release associated with these reactions could be sufficient to cause the evaporation or even the explosion of the organic mantle. In this way, "on grain" synthesised molecules could be injected into the gas phase. Without such a "catastrophic" event, the desorption of complex molecules would never occur, given the very low average temperature of the grains (10–50 K). Nevertheless, it must be pointed out that "hot grains" can exist and that, in addition, other grain spallation processes have been suggested. Interstellar chemistry remains an extremely difficult field, in which generalisations are dangerous. Many different kinds of clouds exist, and they are associated with different kinds of chemistry. The distinction between diffuse, dense and dark, circumstellar clouds has been introduced to point out these differences. The nature and the role of dust particles also varies from one kind of cloud to another. People interested in this fascinating subject might like to read some specialised articles on the topic, such as the paper by Winnewisser [99], or the excellent book by Duley and Williams [13].

To come back to the nature of dust particles, it is now accepted by many authors that they are probably covered with an organic mantle containing elemental carbon and complex organic molecules. Sagan and Khare [104] have been able to show that polymers synthesised by UV radiation or electric discharges in a mixture containing CH_4, C_2H_6, NH_3 and H_2S are stable to about 1000 K, and they conclude that organic polymers are major components of interstellar dust. For obvious reasons, the isotope ratios for carbon, hydrogen and oxygen are unknown as far as the constituent matter of the particles is concerned. Nevertheless, it seems realistic to suggest, as do numerous authors, that grains are constituents of carbonaceous chondrites. To be more precise, some of the carbon (elemental or forming part of the least-volatile organic compounds [105]) found in carbonaceous chondrite had previously been part of the grains and had never been involved in a homogenisation process. The arguments that lead to this very important conclusion are based on the study of noble gases trapped in carbonaceous carrier phases, and these latter constitute approximately 1% of chondritic matter [67]. These carrier phases have already been described. They differ from each other in $\delta^{13}C$ values, in the release temperatures of the noble gases, in the ratios of the isotopes of the noble gases and in their petrographical characteristics; these phases are labelled C_α, C_β and C_δ.

Table 6 gives information on the types of carbon phases found in carbonaceous chondrites, together with some of their characteristics.

At least three exotic noble gas components have been detected in carbonaceous chondrites. The first component is so-called carbonaceous chondrite fission xenon, which is enriched in the heavy and light isotopes of this remarkable element with nine stable isotopes. The carrier carbon phase is characterised by a $\delta^{13}C$ = $-38^o/_{oo}$, and is called carbon-δ. The second component is s-process xenon, which is enriched in even-numbered middle isotopes. The carrier carbon phase is characterised by a $\delta = +1100\%o$, and is called carbon-β. The third component is neon-E(L),

Table 6. Types of carbon in carbonaceous chondrites

Type of carbon	Types of chondrites*	Concentration of C in ppm	δ^{13}C PDB in ‰	Noble gases or other markers; characteristic ratios	Origin
Organic	C1, C2	30000	− 22	Solar values	Protosolar nebula
Elemental	C3	2000	− 22	Solar values	Protosolar nebula
Carbonates	C1, some in C2	2000	+ 72 to + 47	Solar values	Protosolar nebula
Carbon-α	C1, C2, some in C3	5	+ 300	$\dfrac{20_{Ne}}{22_{Ne}} = 0,01 \ (9,8)$	Nova?
Carbon-β	C1, C2, some in C3	5	+1100	$\dfrac{130_{Xe}}{132_{Xe}} = 0,48 \ (0,15)$	Red giant
Carbon-δ	C1, C2, C3	200	− 38	$\dfrac{136_{Xe}}{132_{Xe}} = 0,64 \ (0,32)$	Super nova
Particular fraction of the organic polymer	C1, C2	3000	?	$\dfrac{H}{D} = 1566$	Interstellar molecular cloud

Three types of carbon are local: their properties can be explained by processes in the early solar system. Four types are exotic: Their interstellar origin is revealed by the isotopic composition of carbon or by isotopic anomalies in "markers" that they carry. (From Ref. [98].) Characteristic isotopic ratios of the noble gases in the meteoritic components are given in comparison with their values in the terrestrial atmosphere (in brackets). (From Ref. [67].)

which essentially contains one neon isotope, i.e. ^{22}Ne, associated with a carrier carbon phase with a δ^{13}C of about 300‰. This phase is called carbon-α [67].

The anomalous isotope ratio observed for the noble gases cannot be explained by any chemical process, and isotope mass effects associated with physical processes like diffusion, distillation and absorption–desorption are too small to explain what is observed. On the other hand, the carbon carrier phase is very abnormal, at least for the carbon-β phase. These facts can be explained if we accept that macroscopic amounts of interstellar carbon have survived unchanged, or at least preserved from isotopic exchange with solar system carbon. It is important to observe that a δ^{13}C = +1100‰, which corresponds to ^{12}C/^{13}C = 42, can be compared to the low end of the range observed for carbon in molecular clouds (60 ± 8 or 67 ± 10) [67]. Moreover, the galactic ratio observed is now probably lower than it was 4.5 Gyr ago owing to the stellar production of ^{13}C.

Carbon grains condense when the C/O ratio is higher than 0.9 [67]. Such conditions can only exist in the vicinity of carbon-rich stars like red giants. On the other hand CO, the most abundant carbon-containing molecule in the universe, can be formed at lower C/O ratios and can thus condense from the ejecta of many kinds of stars. Elemental amorphous carbon can be synthesised at a lower temperature from CO, but it is extremely probable that high-temperature condensates directly from gaseous carbon are better crystallised and therefore more resistant than the amorphous carbon just described [97]. Moreover, gases are probably better trapped in a "high-temperature" matrix. All the observations and deductions lead to the conclusion that some of the carbon atoms found in chondritic samples correspond to interstellar

---

* Type means "petrological type" (see p. 87)

grains. No clear evidence can be adduced to substantiate the hypothesis that some of this carbon is not present as elemental carbon but as a component of the carbon skeleton of organic molecules.

As we have already seen, D-enriched macromolecules have a normal $\delta^{13}C$ value, but it is conceivable that the D-enriched component might come from gas-phase interstellar molecules with a normal $\delta^{13}C$ value and an abnormal δD value. On the other hand, some carbon carrier phases could contain both polycyclic aromatic molecules (quasigraphite!) and elemental carbon with all this carbon being abnormal because it forms part of the grains.

Finally, it seems that indirect evidence makes it possible to substantiate the presence of interstellar organic matter in carbonaceous chondrites. It is impossible to prove that unmodified insterstellar molecules are present in meteorites: polymerisation, degradation, cleavage and other reactions probably played a role during accretion, and then during the long life of the parent bodies. Nevertheless, if these reactions did take place, they were unable to cause an isotopic exchange, and this is a severe limitation on any possible subsequent reactions that we might imagine. The following quotation from Swart et al. [67] is directly relevant to this last remark, even if it may correspond to an optimistic view of the real situation: "Interstellar grains have long been among the more elusive astronomical objects, which could be studied only by theoretical or indirect observational methods. It now appears that at least one class of interstellar grains is present in primitive meteorites, and can be studied by direct laboratory techniques".

7.2 The Synthesis of Organic Molecules in the Protosolar Nebula and During the Accretion Process

The protosolar nebula was probably a very complex and dynamic system. In such an environment, the matter constituting the nebula was heavily modified. A characteristic of the protosolar nebula in the collapse stage was obviously its temperature or, to be more precise, the temperature gradient along the radius of the collapsing disc. As we have already pointed out, various cosmothermometers indicate that the temperature where the accretion of the parent bodies of carbonaceous chondrites took place, was around 300–400 K, with the number of H_2 molecules per cubic centimeter around 10^{14}. This observation does not exclude the existence of higher temperatures at a previous stage of the protosolar nebula. Given these conditions, the exact nature of the material available for the accretion of the parent bodies is unknown. Interstellar grains (altered or not), interstellar organic molecules (modified or not) were probably present, as were other small molecules like H_2, CO, CO_2, CH_4, H_2O, N_2 and many others. The alteration of the grains could have led to the oxidation of iron and partial hydration of silicates.

What kind of chemistry can we envisage taking place in such an environment? To answer this question it is important, though not obviously necessary, to ask another question: What was the major carbon-containing molecule in the protosolar nebula? The response "carbon monoxide" seems trivial because we know that CO is the most abundant carbon-containing molecule in the universe. Nevertheless, at 300–400 K and under the pressures prevailing in the protosolar nebula, the mixture

CO—H_2 was no longer thermodynamically stable, as was first noted by Urey in 1953 and discussed in detail in Ref. [1]. Is it sufficient to conclude that CH_4 was the major molecule? Certainly not, because the conversion of CO—H_2 into CH_4 requires catalysts in order to proceed at a rate comparable to the life of the protosolar nebula. These catalysts could be serpentine or magnetite, formed at about 400 K through the reaction of water on olivine or iron and therefore present during the accreation phase. Nevertheless, the conversion factor remains totally unknown. Given these conditions, it is difficult to conclude that CO was no longer present in the protosolar nebula. It is therefore necessary to consider CO/H_2 as a possible starting mixture for subsequent syntheses. Of course, it is impossible to reject CH_4 or even CO_2 as other starting molecules, since the protosolar nebula was probably far from a state of chemical equilibrium.

Starting from H_2, CO, CO_2 and CH_4, how can we synthesise complex molecules like hydrocarbons? The reaction between CO and H_2 in the presence of appropriate catalysts is well known in organic chemistry: it is the Fischer-Tropsch synthesis, which is a possible cosmochemical pathway leading to the formation of complex hydrocarbons. The minerals such as phyllosilicates and magnetite present in the protosolar nebula are efficient catalysts in the performance of Fischer-Tropsch-type syntheses (FTT). This has been clearly demonstrated by numerous authors, including Hayatsu and Anders [1], who have greatly contributed to this field and have given a clear insight into the possible part played by FTT in cosmochemistry. It is important to remember that the FTT reaction is strongly exergonic at 298 K.

$$10 \, CO \, (g) + 21 \, H_2 \, (g) \rightleftharpoons C_{10}H_{22} \, (g) + 10 \, H_2O \, (g)$$
$$\Delta G°(298 \, K) = -879.7 \, kJ \cdot mol^{-1}$$

Simulation experiments have been performed by different research groups, including ourselves. One piece of clear-cut evidence deriving from our work is the fact that the catalytic effects of phyllosilicates like montmorillonite are associated with the preheating of the catalyst at a temperature of 773 K. This preheating is associated with a loss of water of hydration [106] and therefore with a marked alteration in the lamellar structure of the clay. As far as we know, this fact has never been pointed out before, and could be significant if we knew that silicates themselves, such as olivine, had no catalytical effect [1]. It is almost certain that silicates existed in the protosolar nebula, and the presence of phyllosilicates during the late stages of accretion is almost sure.

The kinetic isotope effect observed for the FTT synthesis by Lancet and Anders [71] on CO_2, CH_4, C_{2+} (which means ethane and heavier hydrocarbons) and a wax fraction recovered from the catalyst, is not in disagreement with the experimental values (comparison between Fig. 2, I and II A). The same group [1] have pointed out that in the case of FTT in the presence of NH_3, the amino acid data ($\delta^{13}C$ as high as $+44\%_0$) would not be inconsistent with the experimental values either (see Fig. 1).

It remains true that the subtle signature that corresponds to the $\delta^{13}C$ has never been completely exploited due to the lack, until recently, of very precise $\delta^{13}C$ value measurements for homologous derivatives (cf. Fig. 2 I). Work is in progress in our laboratory to demonstrate clearly that, at least for hydrocarbons, FTT is really the best candidate. The endogenous polymeric matter (which coexists with interstellar polymeric matter) might be the result of a FTT synthesis, as shown by Hayatsu et al. [35], although the isotopic fractionation of ^{15}N measured in the FTT polymer relative to the starting material (NH_3) is too small (3‰) to account for the nitrogen isotopic composition of the organics in CI and CM carbonaceous chondrites [107].

Other pathways are in fact possible, and under this heading the Urey-Miller-type (UMT) synthesis is frequently mentioned. In the UMT synthesis, the initial molecules are highly reduced (CH_4, H_2O, NH_3). In the case of hydrocarbons, the reaction is given by

$$10 \, CH_4 \, (g) \rightleftharpoons C_{10}H_{22} \, (g) + 9 \, H_2 \, (g)$$

$$\Delta G°(298 \, K) = +542.2 \, kJ \cdot mol^{-1}$$

This reaction is strongly endergonic and requires a source of energy. A priori electric discharges, radioactive elements (like the extinct ^{26}Al) and even UV or visible light are good candidates for the energy source [108]. When applied to the synthesis of hydrocarbons, UMT does not give $\delta^{13}C$ as being in agreement with the experimental results [1]. Nevertheless, more recently Des Marais et al. [72] reported an important isotopic fractionation (see Fig. 2 IB) that does not correlate so badly with the experimental results. Moreover, UMT in the presence of CH_4, H_2O, N_2 (with traces of NH_3 or NH_4Cl) is a very efficient way to produce aminoacids, and mono- and dicarboxylic acids [44,45,109]. The polymer synthesised in a UMT reaction was enriched in ^{15}N by only 10‰–12‰ relative to the starting material. This fractionation is too small to account for meteoritic data.

As we said, CO_2 could also act as a starting material. Considering that CO_2 is "heavier" (in terms of ^{13}C content) than CO, all the molecules synthesised from CO_2 would also be heavier. This could explain some differences from class to class of organic molecules (see Fig. 1).

This $^{13}C/^{12}C$ ratio difference between carbon at different levels of oxidation now brings us to a brief discussion of a problem already discussed 40 years ago by Urey [83] in relation to carbon compounds of geological interest (CO_3^{2-}, CO_2, CO, $C_{diamond}$) and extended by Craig [84] to the CH_4 case. These authors studied the equilibration between isotopomers like

$$^{12}CH_4 + {}^{13}CO_2 \rightleftharpoons {}^{13}CH_4 + {}^{12}CO_2$$

From well-known relationships in statistical thermodynamics it is possible to estimate the equilibrium constant of such a type of reaction as far as very accurate values for the partition functions are known. In this way, Craig was able to predict a $\delta^{13}C$ value higher for CO_2 than for CH_4, with enrichment being of the order of 60‰ at room temperature. Lancet and Anders [71] were able to prove that the $\delta^{13}C$ difference between CO_3^{2-} and the reduced organic matter in carbonaceous chondrites might be compatible with equilibration temperatures of the order of 265–335 K. Even if these temperatures are not unrealistic in comparison with accretion temperatures, it is not easy to conceive a reaction mechanism leading to this equilibrium. Moreover, equilibria between the isotopomers of reduced organic molecules and calcium carbonates do not exist in sediments [110]. According to these authors, this does not completely preclude an equilibration (by an unknown mechanism) between gaseous molecules at a very early stage of the accretion process. In our view, and with perhaps one exception, i.e. the CO/CO_2 system, equilibration between isotopomers is a very unlikely explanation of the observed $\delta^{13}C$ values.

From this discussion it can be concluded that the $\delta^{13}C$ values observed for the organic molecules synthesised during the accretion process are most probably the result of kinetic isotope effects. Moreover, the synthetic pathways seem to go from small molecules (CO, CO_2, CH_4) to larger molecules and not in the opposite direction. Evidence for this can be found in the experiments performed by Des Marais et al. [72] (Fig. 2 IIC). Starting from a sample of hexane characterised by a particular $\delta^{13}C$ value (indicated by the letter h in the Fig.), these authors were able to obtain methane, ethane, propane and butane by thermal decomposition at 773 K. The $\delta^{13}C$ calues obtained experimentally for the volatile hydrocarbons decrease with the number of carbon atoms, while the reverse trend is observed for hydrocarbons in Murchison (Fig. 2 I).

At the end of this subsection we are forced to conclude that the origin of the extractable organic matter in chondrites remains fundamentally unknown, though there are at least two plausible scenarios: a Fischer-Tropsch-type synthesis or a Urey-Miller-type synthesis.

Moreover, the isomeric distributions of aminoacids and hydroxyacids seem to suggest that radical reactions, must also be taken into account (see paragraph 5.4). Presently, it remains impossible to determine if this kind of processes played a role during the accretion process or later.

7.3 The Synthesis of Organic Molecules at a Late Stage in the Accretion Process, or Even in the Parent Body

It is important to realise that a large portion of the potentially soluble organic matter is extractable only after demineralisation. This observation proves that low molecular-weight organic matter is trapped in mineral matrices probably in interlayer sites or at grain boundaries. The hydration of silicates and the synthesis of non-macromolecular organic matter were probably two strongly interrelated processes.

It is tempting to describe the parent bodies of the future carbonaceous chondrites, once accretion was completed, as cold objects in motion around the sun. Such a description was broadly accepted until recently. The situation has now changed drastically, and the (unknown) parent bodies of CI 1 and CM 2 carbonaceous chondrites are described as objects that have experienced hydrothermal metamorphism and which therefore were not cold and chemically dead during all their long lives. Collisions between asteroids seem not to be exceptional. Some of these collisions lead to the rupture of the asteroids which then become potential parent bodies. Other less dramatic collisions only produce heating. This heating may be an initial step towards hydrothermal metamorphism.

It is conceivable that chemical reactions may have taken place during metamorphism, and it is difficult to find any reasons for excluding organic reactions. In other words, some of the organic molecules detected today in carbonaceous chondrites could be the products of reactions which happened after accretion in the parent body. What kind of reactions? At this level, it is extremely difficult to suggest any reasonable answer. It is tempting to consider reactions essentially involving hydrolysis or the addition of water but the number of these reactions producing some of the molecules detected in chondrites is far from negligible.

8 General Conclusions

As a final conclusion it can be stated that carbonaceous chondrites are very complex objects. They contain organic molecules synthesised under very different physico-chemical conditions and at different periods. It is not realistic to discuss the origin of the organic matter in carbonaceous chondrites. We must accept that these origins are numerous and highly diverse. The extreme variety of molecular structures found in the Murchison meteorite reflects these various origins but the best signature of these multiple origins can be found in the isotopic ratios of hydrogen and carbon and perhaps, in the future, in oxygen.

In this review article, we have emphasised the importance of δ values as a means of probing synthetical pathways. Isotope effects are of prime importance in many fields including obviously geochemistry, cosmochemistry and astrophysics. In recent years, the general knowledge of isotope effects has increased significantly with the demonstration of the amount of non-mass-isotope effects and, more particularly, the magnetic isotope effect.

The great interest, but also the major difficulty, connected with the field reviewed in this article is its interdisciplinary nature. As chemists, we have emphasised the chemical aspects but, as the same time we have also tried to give a general (and probably oversimplified) overview of the subject.

9 References

1. Hayatsu, R., Anders, E., in: Cosmo- and Geochemistry, Topics Curr. Chem., Vol. 99, Springer-Verlag, Berlin, Heidelberg 1981, p. 1
2. Nagy, B.: Carbonaceous Meteorites, Elsevier Scientific Publishing Co., Amsterdam 1975
3. Spielmann, P. E.: Nature *114*, 276 (1924)
4. Mueller, G.: Geochim. Cosmochim. Acta *1*, 1 (1953)
5. Dodd, R. T.: Meteorites. A Petrologic-chemical Synthesis, Cambridge University Press, Cambridge 1981
6. Van Schmus, W. R., Wood, J. A.: Geochim. Cosmochim. Acta *31*, 747 (1967)
7. Wasson, J. T.: Meteorites. Classification and Properties, Minerals and Rocks, Vol. 10, Springer-Verlag, Berlin, Heidelberg 1974
8. McSween, H. Y., Jr.: Rev. Geophys. Space Phys. *17*, 1059 (1979)
9. Wasson, J. T.: Meteorites. Their Record of Early Solar-System History, W. H. Freeman and Company, New York 1985
10. Bunch, T. E., Chang, S.: Geochim. Cosmochim. Acta *44*, 1543 (1980)
11. Clayton, R. N., Grossman, L., Mayeda, T. K.: Science *182*, 485 (1973)
12. Clayton, R. N., Onuma, N., Grossman, L., Mayeda, T. K.: Earth Planet. Sci. Lett. *34*, 209 (1977)
13. Duley, W. W., Williams, D. A.: Interstellar Chemistry, Academic Press, London 1984
14. Larimer, J. W.: Astrophys. Space Sci. *65*, 351 (1979)
15. Cameron, A. G. W.: Sci. Am. *233*, 33 (1975)
16. Reeves, H.: Philos. Trans. R. Soc. London *A303*, 369 (1981)
17. Clayton, R. N.: Philos. Trans. R. Soc. London *A303*, 339 (1981)
18. Fuchs, L. H., Olsen, E., Jensen, K. J.: Smithson. Contrib. Earth Sci. *10*, 1 (1973)
19. Kvenvolden, K., Lawless, J., Pering, K., Peterson, E., Flores, J., Ponnamperuma, C., Kaplan, I. R., Moore, C.: Nature *228*, 923 (1970)

20. Chang, S., Mack, R., Lennon, K.: Lunar Planet. Sci. *9*, 157 (1978)
21. Pering, K. L., Ponnamperuma, C.: Science 173, 237 (1971)
22. Lawless, J. G., Yuen, G. U.: Nature *282*, 396 (1979)
23. Yuen, G., Blair, N., Des Marais, D. J., Chang, S.: Nature *307*, 252 (1984)
24. Lawless, J. G., Zeitman, B., Pereira, W. E., Summons, R. E., Duffield, A. M.: Nature *251*, 40 (1974)
25. Peltzer, E. T., Bada, J. L.: Nature *272*, 443 (1978)
26. Cronin, J. R., Moore, C. B.: Science *172*, 1327 (1971)
27. Pereira, W. E., Summons, R. E., Rindfleisch, T. C., Duffield, A. M., Zeitman, B., Lawless, J. G.: Geochim. Cosmochim. Acta *39*, 163 (1975)
28. Jungclaus, G. A., Yuen, G. U., Moore, C. B.: Meteoritics, *11*, 231 (1976)
29. Jungclaus, G., Cronin, J. R., Moore, C. B., Yuen, G. U.: Nature *261*, 126 (1976)
30. Hayatsu, R., Studier, M. H., Moore, L. P., Anders, E.: Geochim. Cosmochim. Acta *39*, 471 (1975)
31. Stoks, P. G., Schwartz, A. W.: Geochim. Cosmochim. Acta *46*, 309 (1982)
32. Stoks, P. G., Schwartz, A. W.: Nature *282*, 709 (1979)
33. Stoks, P. G., Schwartz, A. W.: Geochim. Cosmochim. Acta *45*, 563 (1981)
34. Whittaker, A. G., Watts, E. J.: Science, *209*, 1512 (1980)
35. Hayatsu, R., Matsuoka, S., Scott, R. G., Studier, M. H., Anders, E.: Geochim. Cosmochim. Acta, *41*, 1325 (1977)
36. Hayatsu, R., Winans, R. E., Scott, R. G., McBeth, R. L., Moore, L. P., Studier, M. H.: Science *207*, 1202 (1980)
37. Studier, M. H., Hayatsu, R., Anders, E.: Geochim. Cosmochim. Acta *36*, 189 (1972)
38. Yuen, G. U., Kvenvolden, K. A.: Nature *246*, 301 (1973)
39. Oró, J., Gibert, J., Lichtenstein, H., Wikstrom, S., Flory, D. A.: Nature *230*, 105 (1971)
40. Kvenvolden, K. A., Lawless, J. G., Ponnamperuma, C.: Proc. Nat. Acad. Sci. USA *68*, 486 (1971)
41. Lawless, J. G.: Geochim. Cosmochim. Acta *37*, 2207 (1973)
42. Cronin, J. R., Gandy, W. E., Pizzarello, S.: J. Mol. Evol. *17*, 265 (1981)
43. Cronin, J. R., Pizzarello, S., Yuen, G. U.: Geochim. Cosmochim. Acta *49*, 2259 (1985)
44. Zeitman, B., Chang, S., Lawless, J. G.: Nature *251*, 42 (1974)
45. Yuen, G. U., Lawless, J. G., Edelson, E. H.: J. Mol. Evol. *17*, 43 (1981)
46. Mizutani, H.: Nature *276*, 738 (1978)
47. Lawless, J. G., Peterson, E.: Origins Life *6*, 3 (1975)
48. Pollock, G. E., Cheng, C-N., Cronin, S. E., Kvenvolden, K. A.: Geochim. Cosmochim. Acta *39*, 1571 (1975)
49. Bonner, W. A., Blair, N. E., Lemmon, R. M.: J. Am. Chem. Soc. *101*, 1049 (1979)
50. Bonner, W. A., Blair, N. E., Lemmon, R. M., Flores, J. J., Pollock, G. E.: Geochim. Cosmochim. Acta *43*, 1841 (1979)
51. Engel, M. H., Nagy, B.: Nature *296*, 837 (1982)
52. Bada, J. L., Cronin, J. R., Ho, M.-S., Kvenvolden, K. A., Lawless, J. G., Miller, S. L., Oró, J., Steinberg, S.: Nature *301*, 494 (1983)
53. Engel, M. H., Nagy, B.: Nature *301*, 496 (1983)
54. Shimoyama, A., Ponnamperuma, C., Yanai, K.: Nature *282*, 394 (1979)
55. Ramakrishna, K.: Diss. Abstr. Int. *B42*, 3160-B (1982)
56. Cronin, J. R., Pizzarello, S., Moore, C. B.: Science *206*, 335 (1979)
57. Kotra, R. K., Shimoyama, A., Ponnamperuma, C., Hare, P. E.: J. Mol. Evol. *13*, 179 (1979)
58. Folsome, C. E., Lawless, J., Romiez, M., Ponnamperuma, C.: Nature *232*, 108 (1971)
59. Hayatsu, R.: Science 146, 1291 (1964)
60. Hayatsu, R., Studier, M. H., Oda, A., Fuse, K., Anders, E.: Geochim. Cosmochim. Acta *32*, 175 (1968)
61. Van der Velden, W., Schwartz, A. W.: Geochim. Cosmochim. Acta *41*, 961 (1977)
62. Stoks, P. G., Schwartz, A. W., in: Origin of Life, Proc. 3rd ISSOL Meeting, Jerusalem, June 22-27, 1980 (ed. Wolman, Y.) D. Reidel Publishing Co., Dordrecht, Holland 1981, p. 59
63. Smith, P. P. K., Buseck, P. R.: Science *216*, 984 (1982)
64. Pillinger, C. T.: Geochim. Cosmochim. Acta *48*, 2739 (1984)

65. Briggs, M. H.: Nature *197*, 1290 (1963)
66. Clayton, R. N.: Science *140*, 192 (1963)
67. Swart, P. K., Grady, M. M., Pillinger, C. T., Lewis, R. S., Anders, E.: Science *220*, 406 (1983)
68. Kerridge, J. F.: Geochim. Cosmochim. Acta *49*, 1707 (1985)
69. Becker, R. H., Epstein, S.: Geochim. Cosmochim. Acta *46*, 97 (1982)
70. Robert, F., Epstein, S.: Geochim. Cosmochim. Acta *46*, 81 (1982)
71. Lancet, M. S., Anders, E.: Science *170*, 980 (1970)
72. Des Marais, D. J., Donchin, J. H., Nehring, N. L., Truesdell, A. H.: Nature *292*, 826 (1981)
73. Kolodny, Y., Kerridge, J. F., Kaplan, I. R.: Earth Planet. Sci. Letters, *46*, 149 (1980)
74. Robert, F., Epstein, S.: Meteoritics *15*, 355 (1980)
75. Smith, J. W., Rigby, D.: Earth Planet. Sci. lett. *54*, 64 (1981)
76. Yang, J., Epstein, S.: Nature, *311*, 544 (1984)
77. Lewis, R. S., Anders, E., Wright, I. P., Norris, S. J., Pillinger, C. T.: Nature *305*, 767 (1983)
78. Thiemens, M. H., Clayton, R. N.: Earth Planet. Sci. Lett. *55*, 363 (1981)
79. Wright, I. P.: Stable isotopic compositions of H, C, N, O and S in meteoritic low temperature condensates, in: Ices in the Solar System (ed. Klinger, J., Benest, D., Dollfus, A., Smoluchowski, R.) D. Reidel Publishing Company, Dordrecht, Holland, p. 221 (1985)
80. Penzias, A. A.: Astrophys. J. *249*, 518 (1981)
81. Clayton, R. N., Mayeda, T. K.: Earth Planet. Sci. Lett. *67*, 151 (1984)
82. Hulston, J. R., Thode, H. G.: J. Geophys. Res., *70*, 3475 (1965)
83. Urey, H. C.: J. Chem. Soc. *1*, 562 (1947)
84. Craig, H.: Geochim. Cosmochim. Acta *3*, 53 (1953)
85. Richet, P., Bottinga, Y., Javoy, M.: Annu. Rev. Earth Planet. Sci. *5*, 65 (1977)
86. Salikhov, K. M., Molin, Yu. N., Sagdeev, R. Z., Buchachenko, A. L.: Spin polarization and magnetic effects in radical reactions, in: Studies in physical and theoretical chemistry, Vol. 22 (ed. Molin, Yu. N.) Elsevier, Amsterdam (1984)
87. Turro, N. J., Kraeutler, B.: Acc. Chem. Res. *13*, 369 (1980)
88. Turro, J., Kraeutler, B.: Magnetic isotope effects, in: Isotopes in organic chemistry, Vol. 6 (ed. Buncel, E., Lee, C. C.) Elsevier, Amsterdam p. 107 (1984)
89. Thiemens, M. H., Clayton, R. N.: Meteoritics *14*, 545 (1979)
90. Arrhenius, G., McCrumb, J. L., Friedman, N.: Astrophys. Space Sci. *65*, 297 (1979)
91. Reisse, J., Mullie, F.: Isotopic Ratios in the Solar System, Cepadues-Editions, Toulouse, p. 107 (1985)
92. Heidenreich III, J. E., Thiemens, M. H.: J. Chem. Phys. *84*, 2129 (1986)
93. Navon, O., Wasserburg, G. J.: Earth Planet. Sci. Lett. *73*, 1 (1985)
94. Esat, T. M., Spear, R. H., Taylor, S. R.: Nature *319*, 576 (1986)
95. Black, D. C., Pepin, R. O.: Earth Planet. Sci. Lett. *6*, 395 (1969)
96. Yang, J., Epstein, S.: Geochim. Cosmochim. Acta *47*, 2199 (1983)
97. Kerridge, J. F.: Earth Planet. Sci. Lett. *64*, 186 (1983)
98. Lewis, R. S., Anders, E.: Sci. Am. *249*, 54 (1983)
99. Winnewisser, G., in: Cosmo- and Geochemistry, Topics Curr. Chem., Vol. 99, Springer-Verlag, Berlin, Heidelberg 1981, p. 39
100. Geiss, J., Reeves, H.: Astron. Astrophys. *93*, 189 (1981)
101. Geiss, J., Reeves, H.: Astron. Astrophys. *18*, 126 (1972)
102. Greenberg, J. M.: Sci. Amer. *250*, 96 (1984)
103. d'Hendecourt, L. B., Allamandola, L. J., Baas, F., Greenberg, J. M.: Astron. Astrophys. *109*, L12 (1982)
104. Sagan, C., Khare, B. N.: Nature *277*, 102 (1979)
105. Anders, E., Hayatsu, R., Studier, M. H.: Astrophys. J. *192*, L101 (1974)
106. Deer, W. A., Howie, R. A., Zussman, J.: Rock-forming minerals, Vol. 3, Sheet Silicates, Longmans, Green and Co., London 1965, p. 226
107. Kung, C.-C., Hayatsu, R., Studier, M. H., Clayton, R. N.: Earth Planet. Sci. Lett. *46*, 141 (1979)
108. Miller, S. L., Urey, H. C., Oró, J.: J. Mol. Evol. *9*, 59 (1976)
109. Wolman, Y., Haverland, W. J., Miller, S. L.: Proc. Nat. Acad. Sci. USA *69*, 809 (1972)
110. Smith, J. W., Kaplan, I. R.: Science *167*, 1367 (1970)

Organic Molecules in Space

Gisbert Winnewisser[1] and Eric Herbst[2]

1 I. Physikalisches Institut, Universität zu Köln, Zülpicher Straße 77, 5000 Köln, FRG
2 Department of Physics, Duke University, Durham, N.C., USA

Table of Contents

Gisbert Winnewisser and Eric Herbst

The chemistry of interstellar molecular clouds is reviewed. Emphasis has been given to recent observational and theoretical advances. At present approximately 70 identified molecular species have been detected in interstellar molecular clouds and circumstellar shells. Identification of the newly detected species rests mainly on the exacting test between interstellar and laboratory spectra. A large number of unidentified interstellar lines indicates the wide chemical variety in synthesizing molecules in a low temperature, low density environment. Although most of the interstellar molecules are organic, (presently 53) some of the most abundant interstellar molecules are inorganic, such as H_2, H_2O, NH_3. The recent detection of the H_2D^+ molecular ion is astro-chemically important, since the existence of the primary ion H_3^+ has been postulated by the ion-molecule gas phase reaction scheme. Amongst the newly detected molecules are additional carbon chain molecules and the first two ring molecules SiCC and C_3H_2.

Recent theoretical gas phase model calculations leading to the observed molecular complexity are discussed together with a critical evaluation of gas phse versus grain syntheses of interstellar molecules. Finally the question of how large interstellar molecules can be is addressed, seen in the light of chemical evolution.

1 Introduction

The 19th century discovery that the same chemical elements that exist on earth occur outside our earth's atmosphere remains one of the greatest achievements of spectroscopy. In the optical wavelength range Fraunhofer discovered a large number of absorption lines in the solar spectrum, which have now become known as the Fraunhofer lines. Shortly thereafter, Bunsen and Kirchhoff demonstrated with their newly discovered spectroscopic analysis method that the origin of the Fraunhofer lines are the absorbing and emitting atoms of the solar chromosphere. These findings constituted a fundamental discovery by mankind that the sun and stars are composed of the same elements as the earth, and that chemical processes take place in cosmic dimensions outside our own planet.

The basic spectroscopic technique applied by these early investigators of matching frequencies or wavelengths between spectral lines from space and those obtained in the laboratory to determine a spectroscopic identification, has remained in use up to the present day. Now, however, both laboratory scientists and astronomers operate with considerably improved accuracy and far beyond the visible wavelength range throughout the entire electromagnetic spectrum, from the long wavelength or microwave end through the infrared region into the ultraviolet and even beyond.

These first discoveries pertained essentially to stellar spectra and it was not until 1904 that the first indication of matter dispersed in between the stars was deduced from the non-participation of the Na—D absorption lines in the periodic Doppler shift of the delta-Orionis close binary (Hartmann 1904). This and later studies indicated that there was a component of matter along the line of sight between binary stars and the earth, which did not participate in the orbital motion of the binary stars around one another. These observations marked the discovery of interstellar matter. In the 20's and 30's it was also realized that the Na and other gaseous atoms are always mixed with sub-micron-sized particles (dust) which scatter and absorb light from stars, causing an extinction of background starlight. Thus the concept of an interstellar medium consisting of both gas and dust emerged.

The late 30's brought a further important step in the investigation of the interstellar medium — the discovery of the first molecular species. In the optical region, the electronic spectra of the diatomic radicals CH, CH^+, and CN, seen in absorption against the continuum spectra of bright background stars, furnished the first evidence that the interstellar medium was not devoid of molecules but contained at least some simple ones. However, the intensities of the molecular spectral peaks seen via optical absorption studies were quite weak compared with the spectra of atoms, indicating that the sources observed in these early studies were not rich in molecules. These sources, now labeled diffuse interstellar clouds, possess very low gas densities (n $\sim 10^2$ cm^{-3}) and are of limited interest chemically.

Today we know however that the interstellar medium contains a startling number of molecules, including those which had hitherto not even been detected in the laboratory. These molecules are now normally detected in the gas phase by radioastronomers via characteristic spectral emission lines at wavelengths sufficiently long to be largely unaffected by the dust particles. Most molecules are detected not in diffuse clouds but in regions of high gas and dust density. These "dense interstellar clouds" contain enough dust to scatter and absorb wavelengths short of the infrared, making optical absorption measurements impossible. Indeed, due to the strong extinction of starlight by intervening dense clouds, much of the plane of our galaxy is hidden from the earth at visible and shorter wavelengths. This is particularly true for the galactic center area which can only be studied by infrared and longer wavelengths.

Dense interstellar clouds are sufficiently cool (T $= 10 - 50$ K) so that normally only low-lying quantum states of molecules can be thermally excited; these states derive from rigid body rotational motion. Spontaneous emission from excited rotational states occurs at wavelengths longer than the infrared (normally referred to as "microwave"). Discrete infrared emission, which occurs from excited vibrational states, has been detected only in localized hot regions of dense clouds where star formation may be occurring. So far, only two very abundant molecules, CO and H_2, have been recorded via this technique, although rotational emission of molecules in excited vibrational states has been seen for other species. A complete discussion of the physical conditions in dense clouds and how they affect the observed spectra is contained in an earlier article in this series (Winnewisser 1981, see also Winnewisser et al. 1979). We note, however, one very important physical condition determined via spectral studies of dense clouds. The gas densities in even the most 'dense' clouds are quite low by terrestrial atmospheric standards and only range up to about 1.0×10^6 cm^{-3}. Even higher densities are probably encountered in regions of maser emission, where densities up to 10^9 cm^{-3} have been postulated.

Molecular clouds surrounding evolved old stars have also been identified as sources of rich chemical content and are known as circumstellar clouds. The cloud surrounding the old evolved carbon star IRC $+$ 10216 has become the prototype of a chemically rich circumstellar molecular cloud due to the many detected molecules. A carbon star contains more carbon than oxygen, which is unusual in the cosmos.

Table 1 summarizes the approximately 70 presently known interstellar and circumstellar species, revealing a chemically rich and complex composition of the interstellar medium not conceived when the first diatomic molecule OH was detected by radioastronomical techniques by Weinreb et al. 1963. Even when the first polyatomic molecule, NH_3, was discovered in 1968 by Cheung et al. the chemical diversity of molecular clouds presently known could hardly be imagined. At the time of writing this article the molecular complexity ranges from diatomics to the thirteen atom species $HC_{10}CN$. The list of molecules in Appendix I contains all isotopic species discovered to date. A high percentage of the molecules on the list have been discovered quite recently, showing the continued vitality of molecular astronomy. For example, within the last two years HCl, SiCC, CCCO, CCCH, the planar ring molecule C_3H_2, CH_3CCCN, CH_3C_4H, $CH_3C_5N(?)$, and $HC_{10}CN$ have been identified. The list is interesting for several reasons: (i) the chemistry revealed by the observed spectra is essentially an organic chemistry in a strongly hydrogen-rich "reducing" environment similar to that postulated by some scientists for the primeval earth, and (ii) some important molecules are still missing. For example, to date no branched chain molecules have been detected, nor has the simplest amino acid glycine. In addition, in order to have a strong rotational spectrum a molecule must possess a permanent dipole moment. Consequently, non-polar molecules such as CO_2 and HCCH can only be observed by their infrared vibrational spectra, but have not yet been detected in this spectral region because of practical difficulties involved in infrared observations such as atmospheric interference and inherently weak signals. There is so much interstellar

Table 1. Interstellar Molecules (June 1985)

2	3	4	5	6	7	8	9	10	11	13
H_2	H_2O	NH_3								
OH	H_2S									
SO	N_2H^+									
SO^+										
SiO	SO_2									
SiS	HNO									
NO	$NaOH$ (?)									
NS	H_2D^+									
HCl										
PN (?)										
CH^+	HCN	H_2CO	HC_3N	CH_3OH	HC_5N	$HCOOCH_3$	HC_7N	CH_3C_5N (?)	HC_9N	$HC_{11}N$
CH	HNC	$HNCO$	C_4H	CH_3CN	CH_3CCH	CH_3C_3N	$(CH_3)_2O$			
CN	C_2H	H_2CS	H_2CNH	CH_3SH	CH_3NH_2		CH_3CH_2OH			
CO	$SiCC$	$HNCS$	H_2C_2O	NH_2CHO	CH_3CHO		CH_3CH_2CN			
CS	HCO	C_3N	NH_2CN	(H_2CCH_2)	H_2CCHCN		CH_3C_4H			
CC	HCO^+	C_3H	$HCOOH$							
	HOC^+ (?)	C_3O	(CH_4)							
	OCS	$HOCO^+$	(SiH_4)							
	HCS^+	$(HCCH)$	C_3H_2							
	$HCNH^+$									

( ): circumstellar identifications
(?): uncertain identifications

123

Gisbert Winnewisser and Eric Herbst

Appendix I Interstellar Molecules

(June 1985)

Chem. formula	name	isotopic species
"inorganic molecules"		
H_2	hydrogen	HD
OH	hydroxyl radical	^{18}OH, ^{17}OH
SiO	silicon monoxide	^{29}SiO, ^{30}SiO
SiS	silicon monosulfide	^{29}SiS, ^{30}SiS, $Si^{34}S$
NO	nitric oxide	
NS	nitric sulfide	
SO	sulfur monoxide	^{34}SO, ^{33}SO, $S^{18}O$
SO^+		
PN(?)	hydrogen chloride	
HCl	water	HDO, $H_2^{18}O$
H_2O	hydrogen sulfide	
H_2S	H_2D^+-ion	
H_2D^+	diazenylium	NND^+, $HN^{15}N^+$, $H^{15}NN^+$
NNH^+	(dinitrogen monohydride ion)	
SO_2	sulfur dioxide	$^{34}SO_2$
HNO	nitroxyl hydride	
NaOH(?)	sodium hydroxide	
NH_3	ammonia	$^{15}NH_3$, NH_2D
"organic molecules"		
CH	methylidyne radical	
CH^+	methylidyne ion	$^{13}CH^+$
CC	carbon	
CN	cyanogen radical	^{13}CN
CO	carbon monoxide	^{13}CO, $C^{18}O$, $C^{17}O$, $^{13}C^{18}O$
CS	carbon monosulfide	^{13}CS, $C^{34}S$, $C^{33}S$
CCH	ethynyl radical	CCD
SiCC	silacyclopropyne	
HCN	hydrogen cyanide	$H^{13}CN$, $HC^{15}N$, DCN
HNC	hydrogen isocyanide	$HN^{13}C$, $H^{15}NC$, DNC
HCO	formyl radical	
HCO^+	formyl ion	$H^{13}CO^+$, $HC^{18}O^+$, $HC^{17}O^+$ DCO^+, $D^{13}CO^+$
HOC^+ (?)	isoformyl ion	
HCS^+	thioformyl ion	
OCS	carbonyl sulfide	$O^{13}CS$, $OC^{34}S$
H_2CO	formaldehyde	$H_2^{13}CO$, $H_2C^{18}O$, HDCO
H_2CS	thioformaldehyde	
HNCO	isocyanic acid	
HNCS	thioisocyanic acid	
CCCN	cyanoethynyl radical	

124

Chem. formula	name	isotopic species
CCCH	C_3H radical	
CCCO	tricarbon monoxide	
$HOCO^+$	$HOCO^+$ ion	
$HCNH^+$ (?)		
H_2CCO	ketene	
CCCCH	butadiynyl	
H_2CNH	methylenimine (methanimine)	$^{13}CH_2NH$
NH_2CN	cyanamide	
HCOOH	formic acid	
HCCCN	cyanoacetylene	$H^{13}CCCN$, $HC^{13}CCN$,
		$HCC^{13}CN$, DCCCN
C_3H_2	cyclopropenylidene	
NH_2CHO	formamide	$NH_2^{13}CHO$
CH_3OH .	methanol (methyl alcohol)	$^{13}CH_3OH$, CH_3OD
CH_3SH	methyl mercaptan	
CH_3CN	acetonitrile (methyl cyanide)	$CH_3^{13}CN$
CH_3CCH	propyne (methyl acetylene)	
CH_3NH_2	methylamine	
CH_3CHO	acetaldehyde	
H_2CCHCN	acrylonitrile (vinyl cyanide)	
HCCCCCN	2,4-pentadiynnitrile	DCCCCCN
	(cyanodiacetylene)	
$HCOOCH_3$	methyl formate	
CH_3CCCN	methyl cyanoacetylene	
	2-butynenitrile	
$(CH_3)_2O$	dimethyl ether	
CH_3CH_2OH	ethanol (ethyl alcohol)	
CH_3CH_2CN	propionitrile (ethyl cyanide)	
CH_3CCCCH	penta-1,3-diyne	
	(methyl diacetylene)	
CH_3C_4CN (?)	cyano-2,4-hexadiyne	
	(2,4-hexadiynnitrile)	
HC_6CN	2,4,6-heptatriynenitrile	
	(cyanohexatriyne, cyanotriacetyl-	
	ene)	
HC_8CN	2,4,6,8-nonatetraynenitrile	
	(cyanooctatetrayne)	
$HC_{10}CN$	2,4,6,8,10-undecapentaynenitrile	

circumstellar molecular identifications (detected at IR wavelengths):

HCCH	acetylene	
H_2CCH_2	ethylene	
CH_4	methane	
SiH_4	silane	

molecular hydrogen, however, that its pure rotational spectrum has been detected via its weak quadrupolar emission. Another reason for non-detection of certain molecules is that their lowest frequency rotational transitions lie in the millimeter region ($\lambda < 1$ mm) which is just now being opened to observation (Fig. I). Finally, there are other selection criteria which influence the entries in the table but have nothing to do with the actual concentrations of the molecules in dense clouds. These criteria include atmospheric interference at key wavelength ranges, varying cross sections for excitation of excited rotational levels, and the number of energy levels per energy interval. Put simply, this last criterion means that an increasing number of energy levels per energy interval (i.e. the energy level density) causes a decrease of the number of molecules in any given rotational level, and consequently it becomes more difficult to observe the species. This problem is especially important for increasingly complex molecules, which possess large numbers of rotational levels, which are in addition usually split into many sublevels due to possible interactions such as internal rotation and inversion motions.

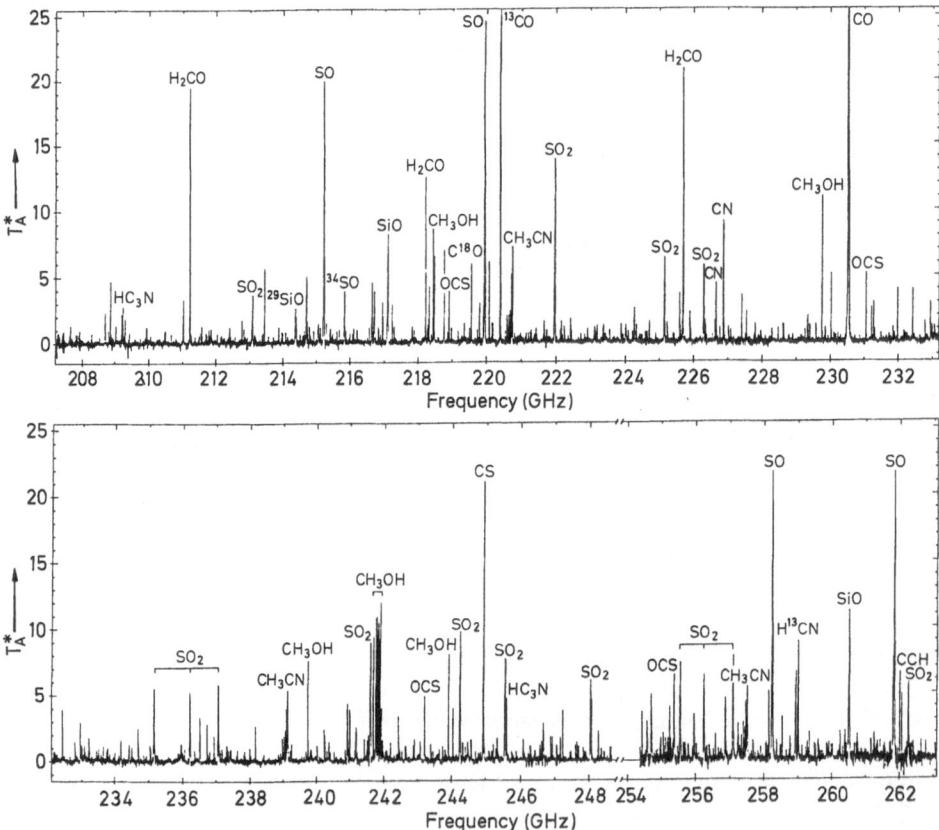

Fig. 1. A low resolution diagram of the millimeter emission spectrum of Orion as seen by the Owens Valley Survey (Blake 1985). Selected intense spectral features are identified on the figure. Most of the weaker features have also been identified

Besides being of interest by themselves, interstellar molecules have become essential tools for astronomers, physicists and chemists interested in the study of the general properties of the interstellar medium. Areas which have been deeply influenced by the observation of interstellar molecules and where substantial new insight into physical and chemical processes have been gained can be placed into four large groups:

(i) Before the advent of studies on interstellar molecules the knowledge of the distribution of neutral matter on a galactic scale, i.e. galactic structure, had been dependent on stellar statistics and since 1951 on measurements of the λ 21 cm hyperfine line of neutral hydrogen. However, since the early 70's it has become evident that the galactic distribution of interstellar molecular clouds can be traced out by the most abundant and widely distributed molecules. Although H_2 is the most abundant interstellar molecule and indeed constitutes 90% of the total gas density, it is of limited use for the surveys of low temperature clouds. This is due to its widely spaced energy levels (the first excited para-H_2 rotational state occurs at over 350 cm^{-1} above the ground state, which puts the lowest rotational quadrupole transition $(J = 2 - 0)$ into the far infrared at 28 µm). Consequently high excitation requirements are needed to populate the higher energy levels. The second most abundant molecule, CO, and its different isotopic species with its low excitation requirements and closely spaced rotational levels, paired with a small permanent electric dipole moment, are perfectly suited to be tracer molecules for interstellar clouds, as long as one assumes that the CO/H_2 ratio is practically constant at 10^{-4}, which occurs if most of the carbon in the clouds is tied up as CO. The lowest rotational transitions of CO, and in particular the $J = 1 \rightarrow 0$ transition at 115 GHz, are therefore extensively being used for general survey purposes, ranging from mapping the details of individual molecular clouds, to surveying the large scale galactic and extragalactic distribution of neutral matter.

(ii) Knowledge of the physics and chemistry of molecular clouds has been gained. Molecular transitions have for the first time yielded precise information on a large variety of cloud parameters, such as extent, size, density, total mass of the clouds, temperature distribution within the cloud, and the various excitation mechanisms. It has been recognized, for example, that certain giant molecular clouds such as the well-studied Orion cloud (OMC-1 or Orion Molecular Cloud 1) have masses approaching 100,000 times that of the sun.

Furthermore, the molecular spectra reveal a plethora of dynamical effects, which are otherwise practically impossible to obtain. Molecular spectra have made close investigations possible of the mechanisms which govern star formation processes. One example is the discovery that in molecular clouds gravitational collapse and subsequent heating leads to the birth of young stars. Associated with star formation are phenomena such as bipolar flow of material in the neighbourhood of young stars, shock fronts emanating from newly born stars and maser emission (H_2O, OH, SiO, CH_3OH) associated with these active regions, as well as the interaction of newly born stars with their maternal molecular clouds. Highly excited rotational molecular lines of many species and vibration-rotation lines of H_2 and CO are the essential tools for probing these hot but confined regions embedded in many molecular clouds.

(iii) Substantial new information has also been obtained concerning the mass loss associated not only with young stars but also with old stars. Stars near the end of their stellar life cycle shed large amounts of stellar processed material (typically 10^{-4} to 10^{-5} M$_\odot$/year) back into space. This is evidenced by the thick circumstellar envelopes covering these old stars — notably the so-called carbon-stars, of which IRC + 10216 is the best studied.

(iv) Finally we have gained totally new insight into the composition and chemistry of molecular clouds, the galaxy as a whole, and other galaxies.

The molecules found to date are composed of the elements H, C, N, O, Si, S, and Cl with the bulk of the molecules containing H, C, N, and O. The light elements H, D, and He are of cosmological origin and are therefore tracers of the early universe. On the other hand the heavier elements C, N, O, . . . are produced in stars by the processes of stellar nucleosynthesis. In addition to the most abundant isotopic forms many stable isotopes such as D, ^{13}C, ^{17}O, ^{18}O, ^{15}N, ^{30}Si, ^{33}S, and ^{34}S have been detected (see Appendix 1). The detailed determination of isotopic ratios — though often beset with formidable difficulties — has become a useful indicator of the chemical evolution of molecular clouds and the past chemical history of the galaxy.

In addition, the last years have revealed that the chemistry in dense interstellar or "molecular" clouds, which is essentially the chemistry of a highly diluted (n $\sim 10^{3-4}$ cm^{-3}) and cold gas (T = 10 — 50 K) coexisting with dust particles, can often show marked differences within the molecular cloud itself: the "hot core chemistry" associated with the "warmer" protostellar environment, embedded in many of the larger ("giant") molecular clouds, produces considerably higher abundances of some interstellar molecules, such as NH_3, due perhaps to less stringent requirements on the types of rapid reactions (see Sect. 3) or evaporation of the outer layers of dust particles, often referred to as "mantles". In the neighborhood of star formation regions supersonic shocks permeate the surrounding molecular material leaving in their wake locally dense and heated gas and a different chemistry now known as "shock chemistry" (see Sect. 4.1).

On the other hand, there are very quiescent dense clouds in which star formation is not a salient feature. A perfect example is the nearby Taurus source (TMC1) which has a relatively uniform cool temperature of 10 K. This source is particularly rich in hydrocarbons and cyanoacetylenes ($HC_{2n}CN$), whereas giant molecular clouds such as OMC1 and the galactic centre source Sgr B2 (Sgr = Sagittarius) are richer in oxygenated organic compounds. On the whole, however, chemical differences between quiescent sources such as TMC1 and the ambient (cooler) portions of giant molecular clouds are not significant (Irvine et al. 1985).

Thus molecular line astronomy has contributed substantially new ideas to many classical fields of astrophysics, but it has also opened up the new discipline of cosmochemistry, the chemistry of a highly diluted, cold gas mixed with solid particles (the interstellar dust) capable of synthesizing molecules often not known to exist under terrestrial conditions. It has remained an intriguing question as to how the interstellar atoms bind together to form fairly complex molecules in this seemingly hostile space environment. In addition, speculation has occurred as to what could be the physical and chemical ties between interstellar molecules and molecules in planetary atmospheres, comets, meteors, and on the surfaces of planets such as our own. In this context it is of wide interest to try to shed more light on the possible connection among the origin of our solar system, its early history, and the molecular cloud from which it was formed. Recent speculation that much of the solar system formed at a cool enough temperature to preserve the interstellar molecules in the surrounding cloud strengthens this connection.

Although amino acids have neither been detected in interstellar clouds, nor within the confined warmer areas of star formation, their discovery is clearly within the scope of present day radio techniques. Though it is feasible to detect the simplest amino acids via radio studies, it will certainly be close to impossible to discover molecules of the complexity of biologically important species such as deoxyribonucleic acid (DNA) or ribonucleic acid (RNA), that govern one of the most important requirements of life — reproduction. But what will be possible is to observe molecules that constitute the functional groups attached to these large biological molecules, but which are amenable to radio astronomical detection. In effect, most of the more elementary functional groups such as $-NH_2$, $-CH_3$, . . . have already been observed as parts of larger molecules. At present, one exception exists: phosphorus-bearing molecules have not yet been detected. From this point of view molecules like PO, PN, HCP, HCCP, PH_3 and others should be searched for to rather low abundance limits. It has however to be mentioned here, that initial searches to detect PN (Morris et al., 1973) and PN, HCP, and PH_3 (Hollis et al. 1980, 1982) remained negative, implying that the chemistry for phosphorus in dense interstellar clouds is either significantly different than for nitrogen, or that phosphorus is severely depleted in the gas phase. A tentative identification has been reported by Sutton et al. 1985. Recent laboratory measurements of reaction rates (Thorne et al. 1984) show that the ion-molecule chemistry of phosphorus is markedly different from that of nitrogen. In fact PH_3 is not readily formed in gas-phase reactions due to the lack of reaction of PH_n^+-ions with molecular hydrogen, H_2. Further speculation on observing large molecules in space is contained in Sections 5 and 6.

Since many basic organic molecules are produced in considerable quantity in the interstellar medium, one is tempted to assume that under more favourable conditions such as higher temperatures and gas densities, biologically important molecules could be easily synthesized. Planetary atmospheres and surfaces are the most likely places for such complex evolution but it must be remembered that the evolution of complex molecules appears to require a reducing atmosphere and it is not at all obvious that a planet such as our own ever possessed such an atmosphere. In any event, in this article in the cosmochemistry series "Topics in Current Chemistry" we confine ourselves to a more modest purpose than an understanding of biological or even pre-biological evolution. Instead, we discuss recent developments concerning the observations and suggested chemical syntheses of intermediate and "large" interstellar organic molecules. The observations are discussed in Section 2, and the chemical formation processes in Sections 3, 4, and 5. In addition, in Section 5 we discuss in the context of chemistry on grain surfaces the limited observations concerning grains and the reason why there is a gas phase at all, given the low temperature of interstellar clouds. Finally, in Section 6, we speculate on how large interstellar molecules can become. This article is written for a general reader with some knowledge of basic chemical and physical processes but with little detailed knowledge of the interstellar medium. A new book on the subject of interstellar chemistry by Duley and Williams (1984) has just appeared and should be of interest to readers of this article.

2 Observations of Organic Molecules in the Gas Phase

Interstellar molecular identifications rest almost entirely on exact comparison with laboratory measured frequencies. Generally, several ways must be pursued to secure a tentative identification, such as measurement of additional rotational transitions, observation of isotopically substituted species, and where possible, observation of fine and/or hyperfine structure. During the late 60's and 70's interstellar discoveries followed in rapid succession due to the availability of a large pool of spectroscopic data, gathered since the early 1950's and which were readily at the disposal of astronomers. However, the pace of new interstellar identifications has slowed some and is now much more closely tied to actual progress in the laboratory both in studying new molecules and new transitions of previously studied species. Within the 15-year history of interstellar molecular spectroscopy via radioastronomy and especially with the recent interstellar frequency searches discussed in Section 2.4, most of the easily accessible molecules and line positions have been checked. In fact, it is the outcome of these searches, that most of the strongest observable interstellar lines have well-known stable molecules as carriers (see Sect. 2.4). However, already in the early 70's exceptions were noticed with the discovery of strong unidentified lines, later assigned to molecular ionic species such as HCO^+ and N_2H^+, radicals such as C_2H, and metastable isomers such as HNC. The assignments were generally made initially by astronomers and only later confirmed by laboratory work.

This general scenario is still being followed today, with several areas left where new discoveries are being and will be made, such as the submillimeter region and adjoining far-infrared region, where many fundamental rotational transitions of light hydrides can be found (for a summary see Winnewisser et al. 1982). In addition, many short-lived molecular intermediates such as radicals and ions have not ceased to be of utmost interest given that unidentified lines continue to be found. Finally the quest for large organic molecules remains an area which has obtained surprisingly little attention, despite facilities well suited for the purpose. The 100 m radiotelescope could take on a leading role in furthering our knowledge substantially. In Section 2.3, we argue that the most important frequency region for large molecule identification is actually the low frequency microwave region. The continuing vitality of molecular astronomy is best illustrated by examples of molecules discovered recently. We divide these species into two groups — stable and unstable molecules — where the adjectives "stable" and "unstable" refer to the terrestrial laboratory environment, and are not very precise terms. Below for example, we identify as "stable" most neutral species in singlet electronic states.

2.1 Recent new Molecular Detections

2.1.1 "Stable" Molecules

HCl

Hydrogen chloride is the first chlorine-bearing interstellar molecule to have been detected. Its lowest rotational transition ($J = 1 \rightarrow 0$) at 625.9 GHz has been observed in the Orion Molecular Cloud (OMC-1) in emission with the Kuiper Airborne Observatory, (Blake, Keene, and Phillips, 1985) since atmospheric opacity at this frequency is high. Although only one rotational transition has been detected, the assignment rests on the observed hyperfine splitting due to the ^{35}Cl nucleus, measured earlier in the laboratory (De Lucia, Helminger and Gordy 1971). Like all other linear molecules in singlet Σ electronic states, HCl has no transitions lower than the J = 1 → 0 line, and can therefore only be seen at submillimeter wavelengths. (λ < 1 mm; ν > 300 GHz).

SiC$_2$

Silicon dicarbide has been identified by Thaddeus et al. (1984) as a circumstellar molecule on the basis of 9 hitherto unassigned millimeter wave lines observed in the late type star IRC + 10216. The molecule is the first ring molecule detected in space, and its rotational spectrum is that of a near prolate asymmetric top with C_{2v} symmetry. The molecule had been detected in the laboratory prior to the interstellar detection by optical laser spectroscopy (Michalopoulous et al. 1984).

C$_3$H$_2$

Cyclopropenylidene is the first truly interstellar ring. Its interstellar and laboratory spectra have been observed by Vrtilek, Thaddeus and Gottlieb (1985). It is an oblate asymmetric top molecule with C_{2v} symmetry. In the laboratory, this planar-ring molecule was produced by a glow discharge in acetylene by Thaddeus et al. (1985a). The interstellar molecule is found in many molecular clouds with widely varying physical conditions. The laboratory spectra of monoisotopically ^{13}C — double-bond substituted C_3H_2 has been reported recently by Bogey and Destombes 1986.

CH$_3$C$_4$H

Methyl diacetylene is the highest member of the methylpolyyne series ($CH_3C_{2n}H$, n = 1, 2, ...) to be detected in interstellar clouds. In the nearby cool cloud TMC1, which has the highest concentration of carbon chain molecules of all interstellar sources discovered to date, CH_3C_4H has been detected via several rotational transitions (Walmsley et al. 1984) (Fig. 2). The astrophysical data together with a recent laboratory value for the dipole moment (Bester et al. 1984) show that the abundance of this species is comparable to that of the cyanopolyynes ($HC_{2n}CN$) as well as to that of methyl acetylene (CH_3C_2H) in this source.

CH$_3$C$_5$N

Methyl cyanodiacetylene has probably been detected by three rotational transitions in the cold dense cloud TMC1 (Snyder et al. 1984). It is somewhat surprising that the inferred abundance ratio of (CH_3C_5N)/(CH_3C_3N) is significantly greater than that of (HC_4CN)/HC_2CN).

HC$_{10}$CN

The largest discovered molecule to date had first been detected in the envelope of the late type star IRC + 10216 and has now also been detected in the interstellar cloud TMC1 (Bell and Matthews 1985). The ratio of the abundances ($HC_{11}N$)/(HC_9N) is 1:4.5 and when combined with abundances of the lower cyanopolyynes one sees a nearly linear decrease of the cyanopolyyne concentration with increasing chain length on a plot of the logarithm of concentration vs. chain length.

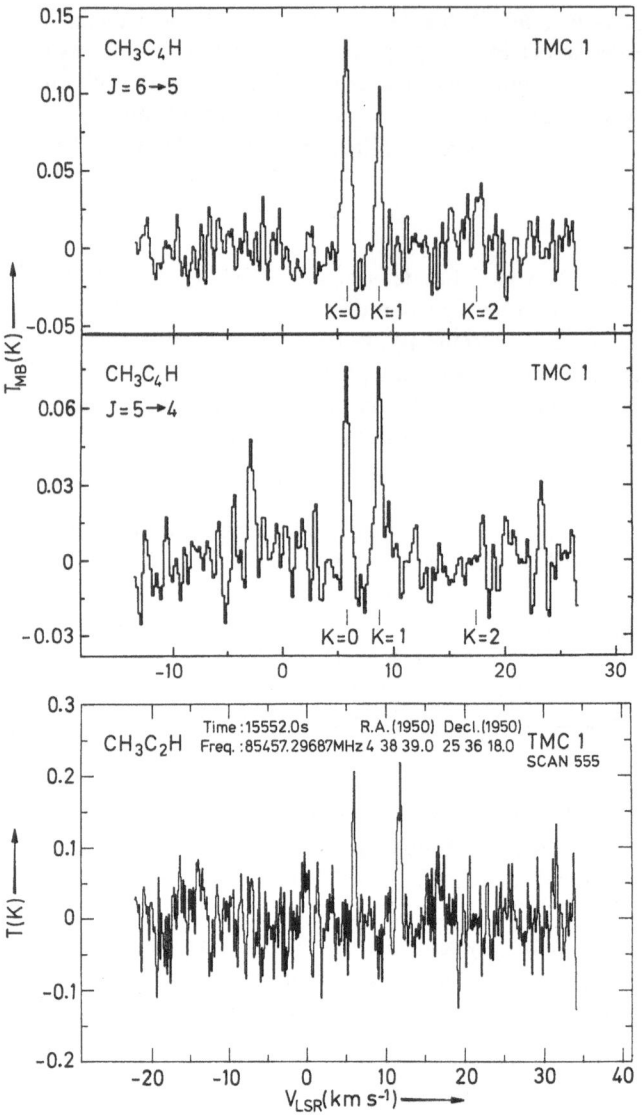

Fig. 2. High resolution spectrum of methyl diacetylene, recently detected with the Effelsberg 100 m-radiotelescope (Walmsley et al. 1984). The bottom trace shows the J = 4 − 3 transition of methyl acetylene, observed with the new Cologne 3 m-radiotelescope. This telescope was located on the roof of the I. Physikalisches Institut, University of Cologne. It has now resumed operation at the High-Alpine-Research-Station Gornergrat-Süd, near Zermatt, Switzerland

2.1.2 "Unstable" Molecules (Radicals and Ions)

The identification of all new radicals has been closely linked to the production of the appropriate species in the laboratory, mainly by glow discharges in acetylene and other molecules. Likewise, the detection of new interstellar ions has depended on laboratory work, typically involving glow discharges of a novel type designed by De Lucia et al. (1983). The newly discovered radicals are C_3N, C_3H, and C_4H reported by Thaddeus and collaborators, all based on the assignment of unidentified interstellar lines in combination with recent laboratory work, and C_3O, the first oxygen bearing carbon chain molecule, which has recently been reported by Matthews et al. (1984). The most recently discovered radicals, C_3H and C_3O, are discussed below as are all recently discovered ions.

C_3H

This radical is a linear carbon chain molecule, discovered in TMC1 and IRC + 10216. The astronomical identification by Thaddeus et al. (1985) has been greatly aided by the laboratory work of Gottlieb et al. (1985), who produced the radical in a glow discharge through a mixture of acetylene, He, and CO. The molecule has a $^2\Pi$ electronic ground state and rotational transitions in both the $^2\Pi_{1/2}$ and $^2\Pi_{3/2}$ spin-orbit substates have been observed. The transition with the lowest frequency observed (a transition across a so-called "lambda doublet") shows a well resolved hyperfine structure. In effect the excellent agreement between the fine structure, rotation, lambda doubling, and hyperfine constants derived from laboratory measurements and astronomical data conclusively confirm the identification.

C_3O

The C_3O radical is an example where recent laboratory work (Brown et al. 1983) has led to a detection, again in TMC1. The molecule can be synthesized in the laboratory by two techniques:
(i) pyrolysis, used by Brown et al. 1983 in their original work to obtain the microwave spectrum and by Klebsch et al. 1985, who recorded the rotational spectrum above 200 GHz.
(ii) glow discharge of carbon suboxide C_3O_2 (Tang et al. 1985).

From monoisotopically substituted versions the structure of the linear molecule has been obtained by Brown et al. (1985).

H_2D^+

With the detection of the molecular ion H_2D^+ in the core region of the molecular cloud NGC2264 by one of its submillimeter wave transitions ($1_{10}-1_{11}$ at 372.4 GHz) using the Kuiper Airborne Observatory (Phillips et al. 1985) the primary ion H_3^+ in the ion-molecule reaction scheme has been found (see Sect. 3, reaction (2)). Once more, this detection rests on recent laboratory work — both Bogey et al. (1984) and Warner et al. (1984) used a magnetically confined glow discharge (De Lucia et al. (1983)) of H_2 and D_2 gases to measure the frequency of the $1_{10}-1_{11}$ transition in the laboratory.

The abundances of deuterated species such as H_2D^+ compared with their hydrogen analogs can be surprisingly large due to the importance of differences in zero-point vibrational energies at low temperature. According to ion-molecule theory (Sect. 3) H_2D^+ is formed via the reaction

$$H_3^+ + HD \rightarrow H_2D^+ + H_2 \tag{2.1}$$

which is exothermic by an energy equivalent to 230 K due to differences in zero-point energy. Therefore the backward reaction at temperatures significantly under 230 K is inhibited by an activation energy of this magnitude. The result is that the H_2D^+ to H_3^+ ratio far exceeds that of HD to H_2 because large amounts of HD can be converted into the trace species H_2D^+.

$HOCO^+$

In 1981 Thaddeus, Guélin, and Linke reported eight new interstellar lines which they assigned to three new "non-terrestrial" molecules. Three of these lines form a triplet at 85 GHz and Thaddeus, Guélin, and Linke (1981) deduced that the three lines of the triplet all belonged to different molecules. One of the triplet lines was immediately assigned to HCS^+ on the basis of new laboratory glow discharge spectra obtained by Gudeman et al. (1981). Furthermore. three lines, which are harmonically related, were boldly assigned to either the linear species $HOCO^+$ or HOCN despite the fact that neither species had been seen in the laboratory! Recently, new laboratory data by Bogey, Demuynk and Destobes (1984) has confirmed the tentative choice of possible assignments made by Thaddeus and collaborators and found $HOCO^+$ to be the carrier. The third member of the 85 GHz triplet is now known to be one of the interstellar lines of the planar ring molecule C_3H_2, cyclopropenylidene (Vrtilek et al. 1985).

Detection of $HOCO^+$ is important because it confirms the existence of the non-polar species CO_2 in dense clouds since ions such as $HOCO^+$ dissociatively recombine in clouds to produce smaller neutrals, viz.,

$$HOCO^+ + e \rightarrow CO_2 + H \tag{2.2}$$

Indeed, use of ion-molecule theory can relate the $HOCO^+$ abundance to that of CO_2. In a similar vein, the recent laboratory measurement of a submillimeter line of H_3O^+ at 307 GHz by Plummer et al. (1985) should enable astronomers to detect this species and deduce the H_2O abundance. Although H_2O has a dipole moment, it has few accessible rotational transition frequencies in the millimeter wave region, and these are heavily obscured by the atmospheric water content. Thus it has not been detected under normal excitation conditions. Submillimeter detection of H_2O at 557 GHz is an alternative possibility to H_3O^+, but will require airborne or even satellite observations due to atmospheric opacity.

HOC^+

HOC^+, the metastable isomer to the well-known HCO^+ molecule, has probably been detected in the galactic centre source Sgr B2 (Woods et al. 1983). This identification rests on only the $J = 1 \rightarrow 0$ transition, which has been measured in the laboratory by Gudeman and Woods (1982). Since this line lies in the galactic centre near several rotational transitions of HCOOH, there remains some doubt as to the proper identification. An unambiguous interstellar identification of HOC^+ would therefore have to await the detection of higher rotational transitions, which have been measured in the laboratory by Blake et al. (1983), or isotopically substituted species. In addition, the HCO^+/HCO^+ abundance ratio of 330 obtained from the observation is at odds witt theoretical determinations (De Frees et al. 1984; Jarrold et al. 1986).

2.2 Determination of Column Densities

One of the important quantitative results that chemists are interested in and which can be derived from interstellar molecular lines is the number of molecules in a given area or volume of space. Under the condition of low optical depth (Winnewisser et al. 1979) the number of molecules contained in a column within an area F (e.g. 1 cm²) along the line of sight to the observer can be derived directly from the observed intensity of an interstellar line. This parameter, called the column density N, is related to the actual concentration n by the equation

$$N = \int_0^L n(x)\,dx \qquad (2.3)$$

where x is the depth into the cloud. For a homogeneous source of length L, the column density is then given by the simple relation $N = nL$. The intensity of an interstellar line is usually expressed in radio astronomy by the concept of a temperature, the "excess brightness" temperature or line temperature T_L. It signifies the temperature a blackbody would have in order to radiate the appropriate amount of energy contained in the line. This line temperature can be derived from the observational results and can then for low optical depth be transformed into a column density. One obtains for the column density N_i in the lower of the two energy levels i and j of a transition (assuming the same nuclear statistical weights for both levels — see for example Winnewisser et al. 1979; Winnewisser et al. 1974) that:

$$N_i = \frac{3kc}{8\pi^3\nu^2}\frac{1}{|\mu_{ij}|^2}\int T_L\,dv \qquad (2.4)$$

where $\int T_L dv$ is the total intensity in the line integrated over frequency. It is often convenient however in deriving column densities from the measured interstellar spectra to represent the total integrated line intensity in velocity space assuming the line width is determined by the Doppler effect. In changing from frequency to velocity space the Doppler relation $dv/v = -dv/c$ is employed. μ_{ij} is the dipole transition moment matrix element connecting the two levels i and j. For a linear molecule undergoing a transition from $J + 1 \leftarrow J$ the dipole moment matrix element μ_{ij} is related to the permanent electric dipole moment μ by (Gordy and Cook, 1984)

$$|\mu_{ij}|^2 = |\mu|^2(J + 1)/(2J + 1) \qquad (2.5)$$

In expression (2.4), ν is the frequency of the transition between levels i and j and k is the Boltzmann constant. Substitution of (2.5) into (2.4) yields

$$N_i = \frac{3k}{8\pi^3 2B}\cdot\frac{(2J + 1)}{(J + 1)^2}\cdot\frac{1}{\mu^2}\int T_L\,dv \sim \frac{3k}{8\pi^3 B}\cdot\frac{1}{(J + 1)}\cdot\frac{1}{\mu^2}\int T_L\,dv \quad (2.6)$$

if level i is characterized by rotational quantum number J and rotational energy $hBJ(J + 1)$, where B is the so-called rotational constant. Determination of the total column density requires a knowledge of the excitation temperature of the species, which

135

is best obtained by measuring the intensities of several lines. For high optical depth, the brightness temperature of the line yields no information about the column density but yields the excitation temperature of the species (Winnewisser et al. 1979).

From an analysis such as described here, the relative abundances of interstellar molecules $N(molecule)/N(H_2)$ in various sources can be determined. Table 2 shows a representative list of abundances in two sources, TMC1 and Orion (OMC1). The relative abundance of a species with respect to H_2 is more uncertain that the absolute column density because it is normally obtained by dividing the absolute column density by the ^{13}CO value and then estimating the ratio of the ^{13}CO to H_2 column densities (Irvine et al. 1985). Thus, one should not assume better than one order of magnitude accuracy in the numbers shown in Table 2. Still, several trends can be discovered in this table:

(i) With respect to H_2, all other molecules are trace constituents of the clouds, with CO the second most abundant molecule at a relative abundance of 10^{-4}.

(ii) Differences exist between the relative abundances of molecules in cooler sources such as TMC1 and warmer ones such as Orion. Still, most relative abundances are quite similar from source to source.

A complete listing of the molecules found in both sources TMC1 and OMC1 is given in Table 3. In Table 4 we list for a selected group of carbon chain molecules column densities for several interstellar sources. Amongst those are the cold clouds TMC1 and L183, the warm sources OMC1 and Sgr B2, and the circumstellar source IRC +10216.

Table 2. Some Observed Molecular Abundances[a]
$(N(H_2) = 10^{22}\ cm^{-2})$

Species	TMC1	Orion "ridge"
H_2	1	1
CO	6×10^{-5}	4×10^{-5}
CCH	8×10^{-9}	8×10^{-9}
NH_3	1×10^{-7}	1×10^{-7}
C_4H	2×10^{-8}	1×10^{-10}
HC_2CN	6×10^{-9}	2×10^{-10}
HC_4CN	1×10^{-8}	2×10^{-11}
C_3H_4	6×10^{-9}	2×10^{-9}
CH_3OH		1×10^{-7}
HCO^+	8×10^{-9}	4×10^{-9}

[a] source: Leung et al. (1984)

Table 3. Molecules Detected in the Orion Molecular Cloud (OMC1) and the Taurus Molecular Cloud (TMC1)

	OMC1	TMC1
Simple hydrides	H_2	
	H_2D^+	
	CH	CH
	OH	OH
	HCl	
	H_2O	

Table 3. (continued)

	OMC1	TMC1
Oxides, sulfides	H_2S	
	NH_3	NH_3
	NNH^+	NNH^+
	CO	CO
	SiO	
	CS	CS
	SO	SO
	SO_2	
	OCS	
	HCO^+	HCO^+
	HCS^+	
		CCCO
Acetylene derivatives	CN	CN
		CCCN
	HCN	HCN
	HNC	HNC
	HCCCN	HCCCN
	HC_4CN	HC_4CN
		CH_3CCCCH
	H_2CCHCN	H_2CCHCN
	HNCO	HNCO
Aldehydes, alcohols, acids ketones,	HHCO	HHCO
ethers, amides, and related	HHCS	HHCS
molecules	HHCCO	
	CH_3CHO	
	$HCOOCH_3$	
	CH_3OH	CH_3OH
	$(CH_3)_2O$	
		HC_6CN
		HC_8CN
		$HC_{10}CN$
	CCH	CCH
		CCCH
		CCCCH
	CH_3CN	CH_3CN
		CH_3CCCN
		$CH_3CCCCCN$ (?)
	CH_3CCH	CH_3CCH

2.3 Frequency Ranges

Radioastronomers have detected interstellar molecules via rotational transitions over a wide range of frequencies from 700 MHz (CH, Ziurys and Turner 1985) to 626 GHz (HCl, Blake et al., 1985). This frequency range (often referred to in its entirety as the "microwave" region) can be subdivided somewhat arbitrarily into three smaller ranges: the microwave (< 50 GHz), the millimeter (50–300 GHz), and the submillimeter (> 300 GHz). Note that a wavelength of 1 mm corresponds to a frequency of 300 GHz, so that radiation of wavelengths considerably larger than 1 mm is still referred to as "millimeter". Each of these regions of the electromagnetic spectrum possesses certain advantages and disadvantages for interstellar molecule identification and analysis. The emission intensity (integrated brightness temperature) in velocity space for a given transition of a molecule in the limit of optical thinness can be calculated from the equation of radiative transfer (Winnewisser et al., 1974 and the previous section) and is approximately proportional to the number of molecules in the emitting state multiplied by the emission frequency. This latter factor clearly favors higher rather than lower frequency observations. The former factor depends on the geometry of the molecule. For a simple linear species, the number of molecules $n(J)$ per cm^{+3} in a state characterized by a Boltzmann distribution and rotational angular momentum quantum number J is given by the expression (see for example Gordy and Cook 1984)

$$n(J) = n \frac{(2J + 1) e^{-E_J/kT}}{\sum\limits_{J=0}^{\infty} (2J + 1) e^{-E_J/kT}} \tag{2.7}$$

Table 4. Distribution of carbon chain molecules (updated from Winnewisser 1981)

Molecule	Cloud type								
	Dark clouds				clouds with assoc. H^+ region			Circumstellar	
	TMC1	L1544	L183	ϱOph.	SgrB2	OriA	W51	IRC + 10216	CRL2688
CN	*			*	*	*	*	*	
HCN	*	*	*	*	*	*	*	*	*
HC$_3$N	*	*	*	*	*	*	*	*	*
HC$_5$N	*	*	*		*			*	
HC$_7$N	*				*			*	
HC$_9$N	*								
C$_2$H	*	*	*		*	*	*	*	*
C$_3$H	*							*	
C$_4$H	*							*	*
C$_3$N	*							*	*
HCCH								*	
C$_2$								*	*
C$_3$									*

Molecular abundances of carbon chain molecules in selected sources (Winnewisser and Walmsley 1979)

Source	Distance d (kpc)	Density $n(H_2)$ (cm^{-3})	Temperature T_K (K)	Molecular column densities (cm^{-2})		
				CN	HCN	HC$_3$N
TMC1	0.1	3×10^4	10	$\sim 10^{13}$	$\sim 10^{14}$	6×10^{13}
L183	0.1	3×10^4	10	$< 3.6 \times 10^{13}$	$\sim 3 \times 10^{12}$	$\sim 10^{12}$
ORIA	0.45	$\sim 10^5$	50–70	$\{ \begin{array}{l} 3 \times 10^{13} \\ -9 \times 10^{14} \end{array}$	10^{15}	2×10^{13}
SGR B2	10	$\sim 10^4 - 10^6$	50	3×10^{14}	—	2×10^{14}
IRC + 10216	0.29		~ 300	1×10^{15}	10^{15}	2×10^{14}

where n is the total concentration of the species, k is the Boltzmann constant, and E_J, the energy of a state characterized by J, is given by the expression

$$E_J = hB\, J(J + 1) \tag{2.8}$$

where B, the rotational constant is determined by the molecular structure (Gordy and Cook, 1984). From the form of equation (2.7), it is clear that n(J) first increases with increasing J until the decaying exponential term dominates. The dipole-allowed transitions of a linear molecule involve the selection rule $\Delta J = \pm 1$, so that the frequency of an allowed $J \to J - 1$ emission line is given by

$$v = E/h = (E_J - E_{J-1})/h = 2BJ \tag{2.9}$$

Since J and v are linearly related, one can see that the dependence of emission intensity on J via n(J) can be expressed in terms of v. Thus the emission intensity is proportional to two factors involving frequency — one a simple v factor and the other a factor which optimizes at a specific frequency. The net result is that the emission intensity optimizes at a specific frequency somewhat higher than the extremum of the n(J) factor. Similar, but more complicated analyses can be performed for non-linear molecules which are either symmetric or asymmetric tops (Gordy and Cook 1984). The important point is that an optimum frequency exists where the emission intensity is a maximum. This intensity, which can be easily calculated by differentiating the formula for the emission intensity, is a function of the rotation constant(s) of the molecule and the temperature of the source. As the size of the molecule increases, its rotation constant decreases, its rotational spectral lines shift to lower frequencies, and its optimum frequency likewise shifts in this direction. As the temperature of a source increases, the population of molecules in higher J states increases via Eq. (2.7). Emission from higher J states results in higher frequencies via eq. (2.9) and the optimum frequency increases in value typically as $T^{1/2}$ (Gordy and Cook 1984; Herbst 1985 e). Thus, higher frequencies are associated with warmer sources and lower frequencies with larger molecules. As a numerical example of the shift of optimum emission frequencies to smaller values as molecular size increases, consider the cyanoacetylenes (cyanopolyynes). Broten et al. (1978) have calculated that at a temperature of 10 K, relevant to cold interstellar clouds such as Taurus (TMC1), the optimum frequencies of HC_5N (HC_4CN), HC_7N, HC_9N, and $HC_{11}N$ are approximately 35 GHz, 25 GHz, 15 GHz, and 10 GHz, respectively. For a warm interstellar cloud such as Orion with a temperature of 50 K, these numbers would be increased by $5^{1/2}$. If one wanted to

HC_5N	HC_7N	HC_9N	C_7H	C_3H	C_4H	C_3N	C_2H_2
7×10^{13}	2×10^{13}	0.3×10^{13}	$\sim 5 \times 10^{13}$	5×10^{12}	$\sim 5 \times 10^{13}$	7×10^{12}	—
—	—	—	$< 6 \times 10^{12}$	—	—	—	—
—	—	—	$\begin{cases} 2 \times 10^{14} \\ 3 \times 10^{15} \end{cases}$	—	—	—	—
2×10^{14}	—	—	—	—	—	—	—
4×10^{14}	10^{14}	—	$\sim 5 \times 10^{14}$	$\sim 3 \times 10^{13}$	$\begin{cases} 4 \times 10^{14} \\ 3 \times 10^{15} \end{cases}$	$\begin{cases} 1 \times 10^{14} \\ 8 \times 10^{14} \end{cases}$	3×10^{19}

study a local star-forming region in Orion at, say, 500 K (assuming one could spatially resolve such a local region) the optimum HC_5N frequency would rise to over 200 GHz. In this discussion we have neglected atmospheric opacity, which becomes serious at frequencies in excess of 250 GHz, and differences in telescope receiver sensitivity as a function of frequency. In general, current technology favors receivers in the frequency range below 300 GHz. We have also used linear molecules as an example because of their simplicity. Analyses with similar qualitative results could be performed for nonlinear species (Herbst 1986) and indicate somewhat higher optimum frequency transitions.

2.4 Recent Surveys

There have been three extensive surveys of molecular emission lines from dense interstellar clouds. The first of these was undertaken by Johansson et al. (1984) and covered the frequency range 72 GHz–91 GHz. This survey was directed at the Orion interstellar cloud as well as the well-known circumstellar source IRC + 10216. For Orion, approximately 190 lines were seen. These lines were assigned to 24 known interstellar molecules, but most of the lines were emitted by a much smaller group of species including methanol, methyl formate, dimethyl ether, and ethyl cyanide. The same pattern was observed to a much greater extent in the more recent survey of Sutton et al. (1985) who looked at Orion in the much higher frequency range of 215–247 GHz. At these frequencies "warm" sources such as Orion are sufficiently hot to excite large numbers of rotational levels of asymmetric rotors such as methanol and methyl formate and these species truly dominate the spectrum. Sutton et al. (1985) identified 517 lines, 70% of which belong to methyl formate, dimethyl ether, ethyl cyanide, methanol, SO_2, and acetonitrile. The identifications were made possible by recent laboratory work at these high frequencies. Like the survey of Johansson et al. (1984), the survey of Sutton et al. (1985) did not immediately result in the identification of any new molecular species. Finally, Cummins et al. (1985) have surveyed the dense cloud located near the galactic center, Sagittarius B2, in the frequency range 70 to 150 GHz and found a similar pattern of a spectrum dominated by known interstellar molecules. Although some unidentified lines were reported in each of the surveys, these constitute only a small percentage of the observed spectral frequencies. Furthermore, it is reasonable to expect that many of the unidentified lines belong to the spectra of previously known molecules but are caused by transitions between highlying energy levels not yet studied in the laboratory. Thus, it seems clear that the frequency ranges studied in the surveys are not most conducive to the discovery of new molecules, especially in "warm" sources where the spectra of previously detected molecules are so dense. Still, these surveys did result in the eventual identification of one new interstellar species — the radical C_3H (first tentatively identified in IRC + 10216 by Johansson et al. 1984) and one unidentified line in the Johansson et al. (1984) survey at 85 GHz belongs to the spectrum of the unusual molecule C_3H_2 (Vrtilek et al. 1985). However, as discussed above, new molecules are more likely to be seen at lower frequencies if the species are complex and at still higher frequencies if the species are light hydrides, whose lowest rotational spectral frequencies lie in the sub-millimeter (Winnewisser et al. 1983). An example of a new complex molecule discovered at lower frequencies is $HC_{10}CN$ (1-cyano-pentadecyne), recently detected at 13.9 GHz in the Taurus molecular cloud by Bell and Matthews (1985) while an example of a higher frequency molecule is HCl, detected at 625.9 GHz by Blake et al. (1985) in Orion, via use of the Kuiper Airborne Observatory to reduce atmospheric opacity.

3 Gas Phase Synthetic Processes Leading to Molecular Complexity

The theory of the formation and depletion of small molecules in the gas phase of dense interstellar clouds has been investigated extensively. For a variety of reasons, including a more detailed understanding of the processes in the laboratory and initial success in predicting the abundance of observed molecular ions, neutrals, and deuterium/ hydrogen chemical fractionation, gas phase synthetic processes have become the dominant explanation of the formation of small interstellar molecules. As originally discussed by Herbst and Klemperer (1973), the low temperature and pressure of interstellar clouds restrict the gas phase chemistry to binary reactions without activation energy. Most reactions between neutral species possess activation energy even if they are exothermic and do not occur under interstellar conditions. Exceptions are reactions involving reactive atoms or free radicals. However, Herbst and Klemperer (1973) found that a more important class of exceptions consist of exothermic ion-molecule reactions. These reactions, studied extensively in the laboratory, are most often found to have no activation energy and therefore little temperature dependence. Indeed, most frequently if some temperature dependence is observed, it is of the inverse variety indicating that ion-molecule reactions are quite rapid at interstellar temperatures. Within the last several years, experimental groups headed by Dunn (J.I.L.A.), Rowe (Meudon), and Arnold (Heidelberg) have managed to measure the rates of ion-molecule reactions at temperatures approaching 10 K and to confirm the rapidity of exothermic ion-molecule reactions at low temperature. For purposes of comparison, it should be noted here that rapid ion-molecule reactions possess rate coefficients on the order of 10^{-9} cm^3 s^{-1}. General articles on the scope of ion-molecule chemistry and the different experimental techniques involved have been published by Huntress (1977) and Smith and Adams (1981). These articles are written within the context of interstellar chemistry. A compendium of results has been circulated by Anicich and Huntress (1984). The large experimental effort in ion-molecule reactions has allowed modelers of interstellar clouds to utilize reactions with either measured reaction rates and products that are sufficiently similar to measured reactions so that their rates and products can be estimated.

The specific mechanisms that produce small interstellar molecules via ion-molecule reactions have been discussed by a large number of investigators since the original paper by Herbst and Klemperer (1973) and a large number of detailed model calculations have been undertaken. For a general discussion of processes, the reader is referred to earlier reviews by us (Winnewisser 1981; Herbst and Klemperer 1976). For the reader who is unfamiliar with gas phase interstellar chemistry, we will briefly consider some of the processes that form and destroy two significant molecules — H_2O and the radical OH.

The initial stage in any in situ model of molecule formation involves the formation of diatomic molecules from atoms. Although there is a gas phase mechanism called radiative association in which two atoms can collide and stick together (see Sect. 3.1), the dominant interstellar molecule H_2 is thought by almost all investigators to be formed on grains when two hydrogen atoms stick to the cold surface, migrate from site to site, and come together to form an adsorbed molecule, which then can evaporate even at cold interstellar temperatures. The dominant problem in grain synthesis of other species, discussed in Section 5.3, is how heavier molecules which cannot evaporate due to stronger adsorption forces, return to the gas phase where they are detected. In any event, once H_2 is formed and ejected into the gas phase, a complex ion-molecule chemistry ensues. The dominant mechanism for ionization is thought to be provided by cosmic rays, which are high energy nuclei that permeate the galaxy. Ultra-violet radiation from external stars and, in giant molecular clouds, from internal stars is not expected to penetrate through the dust significantly, although there is some uncertainty in the degree of UV penetration. The cosmic ray flux is not a large one or else no molecular complexity could be achieved. Rather, it serves to produce a small number of reactive ions. The most important ionization process involves the dominant molecule H_2:

$$H_2 + \text{Cosmic Ray} \rightarrow H_2^+ + e + \text{Cosmic Ray} \tag{3.1}$$

and produces the ion H_2^+ which "immediately" (within a day) reacts with the ubiquitous species H_2 via a well-studied reaction in the laboratory to form the H_3^+ ion:

$$H_2^+ + H_2 \rightarrow H_3^+ + H . \tag{3.2}$$

This ion is not reactive with H_2 but can react with a number of other heavy atoms assumed to be present in initial stages of the cloud. In particular, reaction with oxygen atoms leads eventually to the production of the molecular ion H_3O^+ by a chain of exothermic ion-molecule reactions studied in the laboratory:

$$H_3^+ + O \rightarrow OH^+ + H_2 \tag{3.3}$$

$$OH^+ + H_2 \rightarrow H_2O^+ + H \tag{3.4}$$

$$H_2O^+ + H_2 \rightarrow H_3O^+ + H . \tag{3.5}$$

The protonated water ion, H_3O^+, does not react with H_2 and can eventually recombine with electrons. Ion-electron recombination reactions involving polyatomic ions occur principally via a dissociative mechanism. Although the total rate coefficients of these processes have been measured in several laboratories and found to be quite rapid and to be approximately proportional to the inverse square root of temperature (see, e.g., Mul and McGowern 1980), the neutral fragments have only been detected for one reaction of this type. The situation as regards theoretical treatments is not much better — when the ion and electron come together, the neutral system can normally fragment along one of very many potential surfaces leading to different products in different states of internal excitation. Currently, there exists only one simplified statistical theory of the products of dissociative ion-electron reactions (Herbst 1978). For a number of systems, this theory predicts that dissociative recombination reactions are not specific but form many neutral products. Green and Herbst (1979) have suggested that the most likely bond breakages involve the ejection of a H atom or molecule and most modelers have adopted this approach since the calculation of Herbst (1978) is too detailed for use in models. In the case of H_3O^+, possible products include the neutrals H_2O and OH:

$$H_3O^+ + e \rightarrow H_2O + H; \quad OH + H_2 . \tag{3.6}$$

In general, in ion-molecule models neutral molecules tend to be formed from precursor protonated ions via dissociative ion-recombination reactions.

Once molecules such as H_2O and OH are formed, they are destroyed by either ion-molecule reactions or neutral-neutral reactions, or they are adsorbed onto the grains. The radical OH is a reactive neutral species and it is assumed by most investigators that it can react with neutral atoms such as O and N even at interstellar temperatures. The less reactive species H_2O can only react with ions and, consequently, models predict that there is more H_2O than there is OH. Examples of ion-molecule reactions involving H_2O are

$$HCO^+ + H_2O \rightarrow H_3O^{+\cdot} + CO \tag{3.7}$$

and

$$C^+ + H_2O \rightarrow HCO^+ + H . \tag{3.8}$$

The former reaction does represent a destruction mechanism for water because the protonated water produced can combine with electrons to form OH.

The important interstellar molecule NH_3 is formed via a similar sequence of reactions to (3.3)–(3.6); however, the rate coefficients of some of these reactions are still not well determined at low temperatures. For a recent discussion, see Marquette et al. (1985).

How are detailed models constructed? Let us consider the simple case of an ionic species labelled C^+, which is formed by one ion-molecule reaction and destroyed by another:

$$A^+ + B \xrightarrow{k_1} C^+ + D \tag{3.9}$$

$$C^+ + D \xrightarrow{k_2} \text{Products.} \tag{3.10}$$

The rate of change of the concentration of C^+ — $d[C^+]/dt$ — is given by the differential equation

$$d[C^+]/dt = k_1[A^+][B] - k_2[C^+][D] \tag{3.11}$$

where the symbol "[]" refers to concentration and the k's are rate coefficients. A detailed model is constructed by writing down differential equations of this sort

— labeled kinetic equations — for each species of interest in the model. For example, in the original Herbst and Klemperer (1973) model, there are 100 reactions involving thirty-five species. In contrast, the most recent model of Herbst and Leung (1986) has 1958 reactions and 206 species! In most of the gas phase models published, the only grain process of importance considered is the formation of H_2. The choice of molecular species in the models is heavily determined by the so-called cosmic abundances of the elements, or relative amount of each element. Much data from stars shows that hydrogen is the dominant element, with helium in second place at about 10% of hydrogen by number, and heavier elements such as C, N, and O down by factors of 10^3 to 10^4 from hydrogen. In addition, in most stars the oxygen abundance exceeds the carbon abundance. The situation in the gas phase of interstellar clouds is less clear because some material is needed to make the grains, estimated to constitute 1% of the material. Presumably, the heavy elements are severely depleted from the gas phase. Gas phase modelers have utilized two sets of cosmic abundance ratios — one comes from the diffuse cloud ζ Oph, where optical absorption studies based on starlight that penetrates the cloud indicate elemental depletions of an order of magnitude or so (see, e.g., Prasad and Huntress 1980) and the other contains more severe depletions for elements heavier than oxygen (see, e.g., Graedel, Langer, and Frerking 1982).

Once the cosmic abundance ratios are chosen, one can solve the coupled kinetic equations in a variety of approximations to determine the concentrations of the species in the model as functions of the total gas density. Division of the concentrations by the total gas density utilized in the calculation then yields the relative concentrations or abundances. The simplest approximation is the steady-state treatment, in which the time derivatives of all the concentrations are set equal to zero. In this approximation, the coupled differential equations become coupled algebraic equations and are much easier to solve. This was the approach used by Herbst and Klemperer (1973) and by later investigators such as Mitchell, Ginsburg, and Kuntz (1978). In more recent years, however, improvements in computers and computational methods have permitted modelers to solve the differential equations directly as a function of initial abundances (e.g. atoms). Prasad and Huntress (1980a, b) pioneered this approach and demonstrated that it takes perhaps 10^7 yrs for a cloud to reach steady state assuming that the physical conditions of a cloud remain constant. Once steady state is reached, the results for specific molecules are not different from those calculated earlier via the steady-state approximation if the same reaction set is utilized. Both of these approaches typically although not invariably yield calculated abundances at steady-state in order-of-magnitude agreement with observation for the smaller interstellar molecules.

The approach of Prasad and Huntress (1980a, b) might be called the "chemical time dependent" approach because it utilizes fixed physical conditions. However, the chemical evolution to steady-state takes so long that it may be unwise to maintain fixed physical conditions. Time scales more rapid that the 10^7 years needed to reach chemical steady-state include the grain adsorption time and the free-fall collapse time. Thus, in the simplest approximation, chemical steady-state can never be reached because the gas phase will be adsorbed into the grains and the cloud will have collapsed to form, presumably, a star. The observed facts that clouds do not form stars as rapidly as the free fall model and that they do possess a gas phase, both demonstrate that nature is not simple. In addition, models that ignore these two processes appear to do very well for at least the smaller interstellar molecules (up through four constituent atoms).

One important exception involves the abundance of atomic carbon. In chemical time dependent models, the initial major form of gaseous carbon is chosen to be C^+, because this is the dominant form of carbon in diffuse clouds, those low density regions where the gas is mainly atomic. As the chemistry proceeds, C^+ gradually disappears and C takes its place as the dominant form of carbon. However, as steady state is achieved, the abundance of C drops dramatically, and most of the carbon goes into the form of carbon monoxide, which according to observation is a very abundant interstellar molecule. However, recent observations (Phillips and Huggins 1981; Jaffe et al. 1985) on the fine structure transitions of neutral carbon show that at least some of the well-known interstellar sources have much neutral atomic carbon C, perhaps as much as CO. If the observed atomic carbon is deep inside the clouds, and not just a photodissociation product of CO caused by external stellar ultra-violet radiation penetrating the cloud edges, then there is a disagreement between observation and steady-state theory. The observational evidence that the carbon lies in the central areas of the clouds is controversial (Keene et al. 1985). If we take the view that the carbon is deep inside the cloud, then the simplest way to modify the theoretical results is to assume that the cloud has not reached steady-state. In a typical model (see, for example, Leung, Herbst, and Huebner 1984) the C and CO abundances are roughly equal at a cloud lifetime of approximately 10^5 yrs after the diffuse cloud stage. Postulating such a short lifetime for clouds would not greatly affect many of the other calculated abundances of small molecules. However, the consequences of such short lifetimes include an estimated birthrate for stars which would be significantly too high because all "old" clouds would have become stars, and astronomers favor lifetimes longer by at least one to two orders of magnitude.

With the exception of the atomic carbon problem, the agreement between theory and observation for smaller interstellar molecules is relatively good, especially for those molecules not containing heavy elements such as sulfur and silicon. What is the situation for the larger interstellar molecules; i.e., those greater than or equal to four atoms in size?

3.1 Radiative Association

In the last five to six years, investigators have begun to consider in some detail specific ion-molecule pathways for the production of complex interstellar molecules. The major difficulty in ascertaining these pathways is that as the molecule becomes more complex, the synthetic pathway becomes longer, involving more reactions. As the number of reactions increases, the probability that at least one of these reactions will be either endothermic or, if exothermic, contain an activation energy barrier, increases. Early attempts to write synthetic pathways always seemed to founder because of these two problems. In the nineteen seventies, several groups suggested the possibility of radiative association reactions for mitigating this difficulty (Williams 1972; Herbst and Klemperer 1973; Herbst 1976). Radiative association reactions are processes in which two smaller species come together, and form what chemists would call an unstable "complex". This unstable molecule then either redissociates into reactants or stabilizes itself by emission of sufficient energy in the form of a photon. In order for the two reactants to come together without activation energy, they must either be reactive neutrals (atoms or radicals) or an ion-molecule pair. The process is automatically exothermic unless the two reactant partners cannot form a chemical bond between them. The only possible problem concerns the size of the rate coefficient. After all, the idea of two species "sticking together" does not appear intuitive. Indeed, verification of the radiative association idea in the laboratory was and is difficult because the process only occurs at low pressures. However, at higher pressures, more easily useable in the laboratory, a variant of this process occurs. The variant is labeled three body association and has been studied by a number of investigators in the laboratory, especially for ion-molecule reactions (Adams and Smith 1981; Bohme and Raksit 1985). In this process, the complex is stabilized by collision with a third body, normally helium. Ion-molecule three body association reactions typically possess rate coefficients with strong inverse dependences on temperature (T^{-n}, with $n > 1$), leading to the presumption that radiative association reaction rate coefficients possess the same or similar dependences on temperature. In addition, three body reactions become much more rapid at any given temperature as the size of the reactant partners increases. Thus, three body reactions involving the formation of a diatomic molecule from two atoms are slow whereas reactions involving, for example, the formation of a seven atom molecule can be many orders of magnitude more rapid. Presumably, radiative association reactions are similar in this regard because the larger the complex formed, the longer its lifetime against unimolecular dissociation, and the greater its chance of being stabilized by one method or another.

The last decade has witnessed an intense interest in the theory of radiative association rate coefficients because of the possible importance of the reactions in the interstellar medium and because of the difficulty of measuring these reactions in the laboratory. Several theories have been proposed; these are all directed toward systems of at least three or four atoms and utilize statistical approximations to the exact quantum mechanical treatment. The utility of these treatments can be partially gauged by using them to calculate three body rate coefficients which can be compared with laboratory measurements. In order to explain these theories briefly, it would be helpful to write down equations for the mechanism of association reactions. Consider two species A^+ and B that come together with bimolecular rate coefficient k_1 to form a complex AB^{+*} which can then be stabilized radiatively with rate coefficient k_r, be stabilized collisionally with helium with rate coefficient k_{coll}, or redissociate with rate coefficient k_{-1}:

$$A^+ + B \underset{k_{-1}}{\overset{k_1}{\rightleftarrows}} AB^{+*} \tag{3.12}$$

$$AB^{+*} \overset{k_r}{\longrightarrow} AB^+ + h\nu \tag{3.13}$$

$$AB^{+*} + He \overset{k_{coll}}{\longrightarrow} AB^+ + He. \tag{3.14}$$

At steady-state, the rate coefficient for radiative association in the low pressure limit is given by the equation

$$k_{ra} = \frac{k_1 k_r}{k_{-1} + k_r} \tag{3.15}$$

whereas the rate coefficient for three body association at pressures sufficiently large that radiative association is unimportant is given by the equation

$$k_{3b} = \frac{k_1 k_{coll}}{k_{-1} + k_{coll}[He]} . \tag{3.16}$$

Equation (3.15 is derived in the following manner. The rate of formation of AB^+ via radiative association is given by the rate law

$$d[AB^+]/dt = k_{ra}[A^+] [B] = k_r[AB^{+*}] \tag{3.16a}$$

where "[ ]" signifies concentration. The rate of change of $[AB^{+*}]$ is in turn given by the expression

$$d[AB^{+*}]/dt = k_1[A^+] [B] - (k_{-1} + k_r) [AB^{+*}] \tag{3.16b}$$

where collisional stabilization of the complex AB^{+*} has been neglected. Setting the time derivative in Eq. (3.16b) equal to zero (the steady-state approximation) yields the result that

$$[AB^{+*}] = k_1[A^+] [B]/(k_{-1} + k_r) . \tag{3.16c}$$

Finally, the result for $[AB^{+*}]$ can be inserted into (3.16a) to yield the rate law

$$d[AB^+]/dt = k_{ra}[A^+] [B] = k_1 k_r[A^+] [B]/(k_{-1} + k_r) \tag{3.16d}$$

which immediately leads to (3.15). A similar analysis at higher pressures yields k_{3b} in (3.16).

If redissociation into reactants is faster than stabilization, equations (3.15) and (3.16) simplify into a product of k_1/k_{-1} and either k_r or k_{coll}. Under these conditions, to obtain a theory for a total association rate coefficient, one must calculate both k_1/k_{-1} and k_r or k_{coll}. Three levels of theory have been proposed to calculate k_1/k_{-1}. In the simplest theory, one assumes (Herbst 1980a) that k_1/k_{-1} is given by its thermal equilibrium value. In the next most complicated theory, the thermal equilibrium value is modified to incorporate some of the details of the collision. This approach, which has been called the modified thermal or quasi-thermal treatment, is primarily associated with Bates (1979, 1983; see also Herbst 1980b). Finally, a theory which takes conservation of angular momentum rigorously into account and is capable of treating reactants in specific quantum states has been proposed. This approach, called the phase space theory, is associated mainly with Bowers and co-workers (see, e.g., Bass, Chesnavich, and Bowers 1979). In addition to calculating the value of k_1/k_{-1}, one must also calculate the stabilization rate coefficient. Estimates of k_{coll} range from 10–100% of the collisional value, depending on the temperature (Bates 1979; Herbst 1982a). The latest estimates of k_r indicate a value of approximately 10^3 s^{-1}, caused by infra-red emission between excited quasi-continuous and bound vibrational states (Herbst 1985a).

The modified thermal and phase space theories reproduce most three body association data equally well, including the inverse temperature dependence of the rate coefficient (Herbst 1981; Adams and Smith 1981), and are capable of reproducing experimental rate coefficients to within an order of magnitude (Bates 1983; Bass, Chesnavich, and Bowers 1979; Herbst 1985b). They should therefore be this accurate for radiative association rate coefficients if k_r is treated correctly.

Modified thermal (Bates 1983) or phase space (Herbst 1985c) calculations of radiative association rates indicate, as expected, an inverse temperature dependence and a direct dependence on the complexity of the reaction partners. Thus, if theory is to be believed, the importance of radiative association is enhanced by complex molecules reacting in cold clouds. Let us consider two important examples in the synthesis of interstellar methane (Huntress and Mitchell 1979). Although methane can only be observed with difficulty via radioastronomical methods (by centrifugal distortion induced rotational transitions) because it does not possess a permanent dipole moment, its synthesis is an important one because methane is a precursor to more complex hydrocarbons which can be and have been detected. This synthesis can proceed via the following series of normal and radiative association reactions, most of which have been studied in the laboratory:

$$C^+ + H_2 \rightarrow CH_2^+ + h\nu \tag{3.17}$$

$$C + H_3^+ \rightarrow CH^+ + H_2 \tag{3.18}$$

$$CH^+ + H_2 \rightarrow CH_2^+ + H \tag{3.19}$$

$$CH_2^+ + H_2 \rightarrow CH_3^+ + H \tag{3.20}$$

$$CH_3^+ + H_2 \rightarrow CH_5^+ + h\nu \tag{3.21}$$

$$CH_5^+ + e \rightarrow CH_4 + H \; ; \quad CH_3 + H_2 \, . \tag{3.22}$$

The two radiative association reactions are needed because the normal ion-molecule hydrogenation reactions involving C^+ and CH_3^+ are endothermic. These two reactions are among the few radiative association reactions that have been studied in the laboratory. The $C^+ + H_2$ system has been studied in an ion trap at 10 K and measured to have a rate coefficient below $3 \times 10^{-15} \text{ cm}^3 \text{ s}^{-1}$ (Luine and Dunn 1985). The theoretical value (Herbst 1982b) is $9 \times 10^{-16} \text{ cm}^3 \text{ s}^{-1}$. Barlow, Dunn, and Schauer (1984) have measured the rate coefficient of the radiative association between $CH_3^+ + H_2$ to be approximately $1 \times 10^{-13} \text{ cm}^3 \text{ s}^{-1}$ at 13 K in their ion trap; this value is in order of magnitude agreement with theory (Bates 1983; Herbst 1985a). Note that the size of the radiative association rate coefficient is larger for the more complex ionic reactant and that even for this case it is still four orders of magnitude smaller than the collision rate coefficient of 10^{-9} cm^3 s^{-1}. Some larger radiative association rate coefficients will be discussed below in connection with the syntheses of other organic molecules. It should be noted that even small radiative association rate coefficients are important if one of the reactants is molecular hydrogen because this molecule is so abundant.

3.2 Hydrocarbon Chemistry

Now that we have seen how methane can be formed from atomic constituents, let us consider how more complex hydrocarbons can be produced. We note at the outset that methane is efficiently synthesized from C^+ and C, both of which are more abundant at early stages of the cloud chemistry than at steady state, at which time most of the carbon is in the form of CO. It should not be surprising therefore that, as shall be discussed below, the calculated abundances of methane and species formed from methane are found to peak at times well before steady state conditions are achieved.

There are two major routes to hydrocarbon production via gas phase ion-molecule reactions. These routes can be labeled "C^+ or C insertion" (see Herbst, Adams, and Smith 1983, 1984) and "condensation" (see Mitchell and Huntress 1979, Walmsley et al. 1979, Winnewisser 1981, or Schiff and Bohme 1979). The insertion route for two-carbon hydrocarbons proceeds by reactions such as the following (Herbst, Adams, and Smith 1983, 1984):

$$C^+ + CH_4 \rightarrow C_2H_2^+ + H_2 \; ; \quad C_2H_3^+ + H \tag{3.23}$$

$$C + CH_3^+ \rightarrow C_2H_2^+ + H \tag{3.24}$$

$$C_2H_2^+ + H_2 \rightarrow C_2H_4^+ + h\nu \tag{3.25}$$

$$C_2H_2^+ + e \rightarrow C_2H + H \tag{3.26}$$

$$C_2H_3^+ + e \rightarrow C_2H_2 + H \; ; \quad C_2H + H_2 \tag{3.27}$$

where the first reaction has been studied in the laboratory and the radiative association has been studied indirectly via three body association. Note that further hydrogenation of $C_2H_3^+$ does not occur via normal or association reactions; both have been shown not to occur in the laboratory due to reaction endothermicities and/or activation energy barriers (Herbst, Adams, and Smith 1983). Thus the insertion route cannot produce any hydrocarbon less unsaturated than acetylene. Unfortunately, acetylene, without a permanent dipole moment, cannot be detected via radioastronomy. The radical C_2H, synthesized via reactions (3.26) and (3.27), is a well-known interstellar molecule, however.

The more hydrogenated two-carbon atom hydrocarbons can be synthesized via condensation reactions between hydrocarbon ions and neutrals such as the well-studied reaction

$$CH_3^+ + CH_4 \rightarrow C_2H_5^+ + H_2 \tag{3.28}$$

followed by

$$C_2H_5^+ + e \rightarrow C_2H_4 + H \; ; \quad C_2H_3 + H_2 \; . \tag{3.29}$$

In general, the condensation reaction route is somewhat slower than the insertion route according to current models (viz., Herbst, Adams, and Smith 1984; Herbst 1983; Leung, Herbst, and Huebner 1984) so that ion-molecule treatments typically predict the unsaturated hydrocarbons to be more abundant than the more saturated ones. So far, this prediction appears to be born out by observation — the observed interstellar hydrocarbons are C_2H, C_3H, C_3H_2, C_3H_4 (methyl acetylene), C_4H, and CH_3C_4H, none of which approaches a high degree of saturation.

The insertion route to three-carbon-atom hydrocarbons proceeds via reactions such as:

$$C^+ + C_2H_2 \rightarrow C_3H^+ + H \tag{3.30}$$

$$C_3H^+ + H_2 \rightarrow C_3H_3^+ + h\nu \tag{3.31}$$

$$C_3H_3^+ + e \rightarrow C_3H_2 + H \; ; \quad C_3H + H \tag{3.32}$$

where further hydrogenation does not occur via any type of reaction (Herbst, Adams, Smith 1983, 1984). Reaction (3.30) has been studied in the laboratory and reaction (3.31) has been studied indirectly via three body association. The most stable form of the $C_3H_3^+$ ion is a ring (Rosenstock et al. 1977), and it is not surprising that the radical C_3H_2, recently discovered in the laboratory and in space (Vrtilek et al. 1985) is also ring-shaped.

Once again, condensation reactions are needed to produce more saturated species such as methyl acetylene which is produced by reactions such as

$$C_2H_2^+ + CH_4 \rightarrow C_3H_5^+ + H \tag{3.33}$$

followed by

$$C_3H_5^+ + e \rightarrow C_3H_4 + H \; , \tag{3.34}$$

where reaction (3.33) has been studied in the laboratory. The situation for four-carbon-atom hydrocarbons is similar. Interstellar C_4H can be formed both by insertion and condensation reactions:

$$C + C_3H_3^+ \rightarrow C_4H_2^+ + H \tag{3.35}$$

$$C_2H_2^+ + C_2H_2 \rightarrow C_4H_2^+ + H_2 \tag{3.36}$$

$$C_4H_2^+ + e \rightarrow C_4H + H \; . \tag{3.37}$$

while condensation reactions are the only mechanism for producing more saturated hydrocarbons. As hydrocarbons become larger than this, the amount of available laboratory and thermodynamic information becomes quite meagre and it is diffcult to determine the correct reaction pathways towards increasing complexity. However, we can speculate that the complex interstellar hydrocarbon CH_3C_4H might be formed by condensation reactions such as:

$$C_4H_2^+ + CH_4 \rightarrow C_5H_5^+ + H \tag{3.38}$$

followed by dissociative ion-electron recombination.

In the above hydrocarbon syntheses, many of the important reactions have been studied in the laboratory, albeit mainly at room temperature. Important exceptions are reactions involving atomic carbon which, as we will see in our discussion on detailed results of models, appear to play a significant quantitative role in the syntheses of hydrocarbons and other organic molecules. The extension of ion-molecule syntheses to even larger hydrocarbons than shown above will require additional studies in the laboratory to determine relevant rates and products for both the insertion and condensation pathways.

3.3 Cyanoacetylene (Cyanopolyyne) Chemistry

The observation of a series of cyanoacetylenes in interstellar clouds (C_2CN radical, HC_2CN, HC_4CN, HC_6CN, HC_8CN, $HC_{10}CN$) has evoked intense interest concerning how they are synthesized and how large they can grow under interstellar conditions. A variety of proposals have been suggested for gas phase ion-molecule syntheses of these species. In this regard, it should be noted that in laboratory discharges of acetylene and HCN, the two lowest cyanoacetylenes (HC_2CN, HC_4CN) have been detected (Winnewisser et al. 1979). The appropriate methylcyanoacetylenes, e.g. CH_3CCCN have been produced by discharging methyl acetylene, CH_3CCH, and HCN (Winnewisser et al. 1980), which has lead to the prediction of its interstellar existence. These discharge experiments imply that gas phase ion-molecule reactions can indeed produce at least these species although the exact mechanism is still not understood. Perhaps the first mechanism suggested involved reactions of hydrocarbon ions and HCN (Walmsley et al. 1979; Toelle et al. 1979; Winnewisser and Walmsley 1979; Schiff and Bohme 1979; Huntress and Mitchell 1979). For production of C_2CN and HC_2CN, the reaction sequence is

$$C_2H_2^+ + HCN \rightarrow H_2C_3N^+ + H \tag{3.39}$$

$$H_2C_3N^+ + e \rightarrow HC_2CN + H \; ; \quad C_2CN + H_2 \; . \tag{3.40}$$

Somewhat after this synthesis appeared in the literature, it was criticized by Mitchell, Huntress, and Prasad (1979) who proposed a synthesis involving radiative association reactions between protonated nitrogen-containing ions and hydrocarbon neutrals. In this approach the synthesis of the lowest members of the series proceeds by

$$HCNH^+ + C_2H_2 \rightarrow H_4C_3N^+ + h\nu \tag{3.41}$$

$$H_4C_3N^+ + e \rightarrow HC_2CN + H_2 + H \; ; \quad C_2CN + 2\,H_2 \; . \tag{3.42}$$

Both of these approaches involve collisions of two large species. According to model calculations by one of us (Herbst 1983), neither of these syntheses can reproduce the observed high abundance of HC_3N in sources such as the nearby dense interstellar cloud TMC-1. However, this negative assessment for the first mechanism relies on a laboratory rate coefficient for reaction (3.39) at room temperature that shows the reaction to be slow; slow ion-molecule reactions at room temperature often become more rapid at lower temperature (Rowe et al. 1984). Likewise, the negative assessment for the second mechanism is based on a theoretical calculation for the rate coefficient of the radiative association reaction.

Another, and more likely, possibility for cyanoacetylene synthesis involves reactions between elemental nitrogen in the form of N or N^+ and either ionic or neutral hydrocarbons (Herbst 1983; Suzuki 1983; Millar and Freeman 1984a, b). Models show that much of the cosmic abundance of nitrogen in dense clouds resides in neutral atomic form. Until recently, the calculated abundance of N^+ was quite low. However it has just been realized that the heretofore major destruction mechanism of N^+ — the reaction between N^+ and H_2 — is slightly endothermic and may be unimportant (Luine and Dunn 1984). The synthetic reactions involving N and N^+ have for the most part not been studied in the laboratory at any temperature and are highly speculative. Let us first consider atomic nitrogen reactions. Synthetic routes to HC_2CN and C_2CN include

$$C_3H_3^+ + N \rightarrow H_2C_3N^+ + H \tag{3.43}$$

$$C_3H_2^+ + N \rightarrow HC_3N^+ + H \tag{3.44}$$

$$HC_3N^+ + H_2 \rightarrow H_2C_3N^+ + H \tag{3.45}$$

followed by dissociative recombination

$$H_2C_3N^+ + e \rightarrow HC_2CN + H ; \quad C_2CN + H_2 \tag{3.46}$$

as well as the neutral-neutral reaction

$$C_3H_2 + N \rightarrow HC_2CN + H . \tag{3.47}$$

If these reactions occur with significant rate coefficients, they are sufficiently rapid to reproduce the observed abundances of HC_2CN and C_2CN (Herbst, Adams, and Smith 1984; Leung, Herbst, and Huebner 1984; Herbst and Leung 1986; Millar and Freeman 1984a, b). However, it is far from clear that they are rapid. Reaction (3.43) involves a spin flip from ground state reactants to ground state products and may therefore be slow unless a deep reaction well exists (Federer et al. 1984). In addition, reaction (3.45) is quite slow at room temperature (Knight et al. 1985) and may be slow at interstellar temperatures as well. Thus, an aura of uncertainty exists concerning this synthetic mechanism although it is the most likely of the three presented. An extension of the

N atom synthetic mechanism to HC_4CN has been suggested (Millar and Freeman 1984a, b) with reactions such as

$$C_3H^+ + C_2H_2 \rightarrow C_5H_2^+ + H \tag{3.48}$$

$$C_5H_2^+ + N \rightarrow HC_5N^+ + H \tag{3.49}$$

$$HC_5N^+ + H_2 \rightarrow H_2C_5N^+ + H \tag{3.50}$$

followed by dissociative recombination.

What about N^+ reactions? One possible reaction pathway might be

$$N^+ + C_3H_2 \rightarrow HC_3N^+ + H \tag{3.51}$$

followed by hydrogenation (Reaction 3.45) and dissociative recombination (Reaction 3.46). Reactions such as these should be considered by modelers and by laboratory groups if the abundance of N^+ is calculated to be reasonably high. The determination of the actual N^+ abundance will require an estimate of the rate of the interstellar reaction between N^+ and H_2 which differs from that measured in the laboratory because the laboratory H_2 is "normal" (three parts ortho to one part para) no matter how low the temperature utilized whereas the interstellar H_2 is thermalized and therefore mainly in the lowest ($J = 0$) para state.

In the above discussion we have focused on forming cyanoacetylenes from neutral and ionized atomic nitrogen. Why have we ignored N_2? The atomic nitrogen abundance in dense clouds is normally less than that of molecular nitrogen (Leung et al. 1984); however, the chemical bond in N_2 is so strong that breaking it to form cyanoacetylenes is difficult and most often endothermic.

A scheme for obtaining higher members of the cyanoacetylene series from lower members has been advocated by Bohme (1985) based on some work by Bohme and Raksit (1985). The idea is that C^+ insertion followed by reaction with CH_4 can produce the next higher member of the series. As an example, consider

$$C^+ + HC_2CN \rightarrow C_4N^+ + H \tag{3.52}$$

$$C_4N^+ + CH_4 \rightarrow H_2C_5N^+ + H_2 \tag{3.53}$$

followed by dissociative recombination. The first of these reactions has been measured to be exceedingly rapid in the laboratory and the second is based on analogous reactions. Reaction (3.52), involving an ion and a polar reactant, may be even more rapid at low interstellar temperatures.

Our brief survey of suggested pathways to cyanoacetylenes shows that a variety of mechanisms have been proposed, but that insufficient laboratory data precludes a definitive answer on which if any of the mechanisms are viable.

3.4 Other Complex Molecules

We have so far discussed gas phase syntheses for hydrocarbons and cyanoacetylenes. The hydrocarbons were discussed first because of the relative conceptual simplicity of their synthetic pathways and the cyanoacetylenes were discussed next because of the intense interest in these species and because their syntheses are intimately involved with those of the hydrocarbons. However, there are many different types of organic molecules in the gaseous interstellar medium and we shall attempt to discuss syntheses of a representative sample of these species.

It was suggested by first Smith and Adams (1978) and then Huntress and Mitchell (1979) that radiative association reactions involving the methyl ion CH_3^+ (synthesized via Reaction 3.20) and a host of small neutral species could lead to a variety of more complex ion precursors to organic molecules. The rate coefficients of these radiative association reactions have since been calculated by Bates (1983) and Herbst (1980a, b; 1985a, c). Let us consider some of these reactions and the subsequent dissociative recombination reactions:

$$CH_3^+ + CO \rightarrow CH_3CO^+ + h\nu \tag{3.54}$$

$$CH_3^+ + H_2O \rightarrow CH_3OH_2^+ + h\nu \tag{3.55}$$

$$CH_3^+ + HCN \rightarrow CH_3CNH^+ + h\nu \tag{3.56}$$

$$CH_3^+ + NH_3 \rightarrow CH_3NH_3^+ + h\nu \tag{3.57}$$

$$CH_3^+ + CH_3OH \rightarrow CH_3OCH_4^+ + h\nu \tag{3.58}$$

$$CH_3CO^+ + e \rightarrow CH_2CO \text{ (ketene)} + H \tag{3.59}$$

$$CH_3OH_2^+ + e \rightarrow CH_3OH + H \tag{3.60}$$

$$CH_3CNH^+ + e \rightarrow CH_3CN(CH_3NC) + H \tag{3.61}$$

$$CH_3NH_3^+ + e \rightarrow CH_3NH_2 + H \tag{3.62}$$

$$CH_3OCH_4^+ + e \rightarrow CH_3OCH_3 + H \,. \tag{3.63}$$

The formation of CH_3CNH^+ via reaction (3.56) can lead to both CH_3CN and its metastable isomer CH_3NC. A metastable isomer is a higher energy form which, however, contains an activation energy barrier against isomerization. DeFrees et al. (1985) have shown that the ion produced in (3.56) rapidly isomerizes between the protonated cyanomethane and isocyanomethane forms. Upon reaching thermal equilibrium, the CH_3CNH^+ ion dominates by perhaps a factor of ten in concentration over the CH_3NCH^+ ion due to its lower energy (Fig. 3). Therefore, when the neutral species CH_3NC and CH_3CN are formed via ion-electron dissociative recombination of these precursor ions, cyanomethane is predicted to be dominant by this same factor of ten since stable and metastable species are depleted at roughly equal rates via ion-molecule reactions. This theoretical result is in agreement with the observed upper limit for the CH$_3$NC abundance in TMC-1 (Irvine and Schloerb 1984). Unlike CH$_3$NC, the metastable isomers HNC and HOC$^+$ are known interstellar molecules. Their chemistry has been reviewed by Irvine et al. (1985). Green and Herbst (1979) have listed many more possible metastable isomers for interstellar detection; unfortunately, most of these molecules have yet to be studied in the laboratory. Recently, however, Schwahn et al. (1985) have measured the millimeter spectrum of CH$_2$CCHCN, a metastable isomer of the newly discovered interstellar organic molecule CH$_3$C$_3$N.

One might ask why the CH$_3^+$ ions has been chosen for the special task of precursor to so many well-known interstellar organics. There are two reasons. The ion CH$_3^+$ is relatively abundant in interstellar models because it is only depleted slowly via reaction with hydrogen. In addition, CH$_3^+$ has been measured to undergo three-body association with a wide variety of neutral species including the reactants shown above. It seems reasonable to assert that other relatively abundant hydrocarbon ions will also associate with small neutrals to form a wide variety of organic ion precursors. For example, Herbst, Smith, and Adams (1984) have shown that the relatively abundant ions C$_2$H$_2^+$ and C$_2$H$_3^+$ associate with CO in the laboratory via the three body process. Via the radiative association mechanism in interstellar clouds, these reactions lead to interstellar C$_3$O, a recently discovered interstellar radical:

$$C_2H_2^+ + CO \rightarrow H_2C_3O^+ + h\nu \tag{3.64}$$

$$C_2H_3^+ + CO \rightarrow H_3C_3O^+ + h\nu \tag{3.65}$$

$$H_2C_3O^+ + e \rightarrow C_3O + H_2 \tag{3.66}$$

$$H_3C_3O^+ + e \rightarrow C_3O + H_2 + H . \tag{3.67}$$

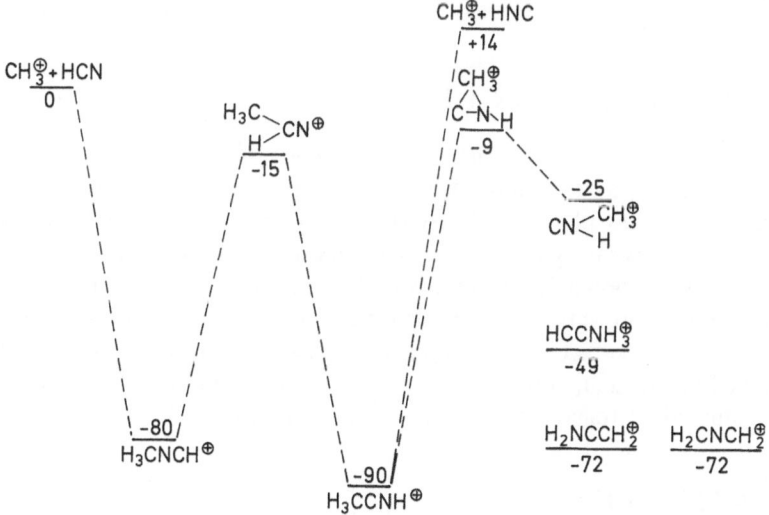

Fig. 3. The potential energy surface of the precursor ions to the isomers CH$_3$NC and CH$_3$CN as calculated by DeFrees et al. (1985). Energies are in kcal mol^{-1}. See text for discussion

Other radiative association reactions leading to formic acid, ethanol, and sundry species are discussed in Leung, Herbst, and Huebner (1984). Calculations of radiative association rate coefficients for ion-molecule systems with large numbers of atoms will be necessary to extend gas phase mechanisms to the syntheses of still larger species. Unfortunately, such calculations are often rendered difficult by the lack of suitable thermodynamic, structural, and vibrational data on the product ions which are needed as input into the calculations. A somewhat easier approach is to estimate the radiative association rate coefficient from higher temperature laboratory three-body rates (Smith et al. 1983). Even so, this approach cannot be used for most reactions of interest involving more complex reactants because of a paucity of laboratory measurements. It is clear that more laboratory work will always be needed!

One should not get the impression that all of the organic molecules in interstellar clouds that are not hydrocarbons or cyanoacetylenes can only be formed by radiative association reactions. Let us consider as counter examples the molecules CH_2CHCN (vinyl cyanide) and CH_3C_2CN. These nitrogen-containing molecules can be formed from hydrocarbon ions and nitrogen atoms (Millar and Freeman 1984a but see the reasons for caution discussed above). Consider for example

$$C_3H_5^+ + N \rightarrow H_4C_3N^+ + H \tag{3.68}$$

$$H_4C_3N^+ + e \rightarrow CH_2CHCN + H \tag{3.69}$$

where the $C_3H_5^+$ ion was synthesized via reaction (3.33). Or consider

$$C_4H_3^+ + N \rightarrow H_2C_4N^+ + H \tag{3.70}$$

$$H_2C_4N^+ + H_2 \rightarrow H_3C_4N^+ + H \tag{3.71}$$

$$H_3C_4N^+ + H_2 \rightarrow H_4C_4N^+ + H \tag{3.72}$$

followed by

$$H_4C_4N^+ + e \rightarrow CH_3C_2CN + H \, . \tag{3.73}$$

Of course, it is unclear whether or not the above hydrogenation reactions are even exothermic. Once more the need for more laboratory work appears pressing.

The above examples of gas phase syntheses should indicate the number of ways in which gas phase processes can synthesize organic molecules. One should remember that many of the reactions shown in our discussion have not been studied in the laboratory and that many more have been studied only at room temperature. Thus, we can expect some surprises and modifications of existing pathways when more reactions are studied at low temperature. One interesting speculation by Rowe et al. (1984) based on some laboratory work over the temperature range 20–560 K on the reaction

$$O_2^+ + CH_4 \rightarrow CH_3O_2^+ + H \tag{3.74}$$

is that many of the ion-molecule reactions that are slow at room temperature become faster as the temperature is reduced. A theoretical model for such reactions has been suggested (Bass et al. 1983) in which the potential surface plays more of an active role than is normally considered to be the case in ion-molecule reactions. The idea is that once the reactants collide to form a complex, the complex cannot easily exit to products but must traverse a transition state barrier. This barrier increases with angular momentum due to centrifugal forces so that high angular momentum collisions have more difficulty in crossing it. The higher the temperature, the larger the collision energies, the larger the range of angular momenta, and the more difficult it is for the high angular momentum complexes to form products.

Despite the present uncertainty in the values of ion-molecule rate coefficients at interstellar temperature, it is appropriate at this point to ask whether or not gas phase pathways produce sufficiently large abundances of the well known organic molecules to agree with observational results. To answer this question, we must discuss the results of the detailed models which contain complex molecules and which have been published.

4 Gas Phase Model Calculations of Complex Molecule Abundances

Model calculations that include at least some of the reactions we have discussed for the syntheses of complex molecules have been performed in the last several years. Both steady-state and chemical time dependent models have been published. Unfortunately, as models include more and more complex species, they become more and more sensitive in their predictions to small changes. As an example, consider two models that in their predictions of the abundances of one-carbon-atom hydrocarbons differ by a factor of 3. This factor is not considered to be a major one in the field of interstellar chemistry. However, since the two-carbon-atom hydrocarbons are formed by reactions between one-carbon atom species, the model will differ in their predictions for the abundances of the larger hydrocarbons by a factor of 9. As one can easily discern, the situation becomes worse as the size of the hydrocarbons increase. Given this extreme sensitivity, modelers should attempt to make sure that at each stage of molecular complexity, they consider all depletion mechanisms and do not overestimate the abundances of the complex molecules that are intermediates in the formation of still more complex species. Unless this is done, models can become in our view overly optimistic about the growth of complexity in the interstellar medium.

Let us first discuss the steady-state models. Herbst (1983) and Herbst, Adams, and Smith (1984) have published what they term "semi-detailed" models of the chemistry of selected complex molecules. These authors do not attempt to calculate the abundances of all molecules but rather fix the abundances of the smaller species at either observed or previously calculated values and then calculate the abundances of more complex species. In this way, the authors feel that model uncertainties in the smaller species will not affect calculations of the larger ones. Using many of the reactions discussed above, the authors find that the abundances of hydrocarbons such as C_3H_4 (methyl acetylene) and C_4H, as well as cyanoacetylenes such as C_2CN and HC_2CN are calculated to be much too low unless the abundance of atomic carbon is set at a rather high level. Presumably a high abundance of atomic carbon aids in the initial step of hydrocarbon synthesis (see React. 3.18) as well as in insertion reactions which then results in higher abundances of more complex species. This fascinating result is not in disagreement with observation, since atomic carbon has been seen in several interstellar sources, but is in disagreement with detailed steady-state predictions, in which most of the atomic carbon has been converted into CO.

On the other hand, Millar and Freeman (1984a, b) have utilized a more detailed steady-state approach to the chemistry of complex hydrocarbons and nitrogen-containing species. These authors find some agreement between theory and observation without a large abundance of atomic carbon. For example, for the interstellar cloud TMC-1, Millar and Freeman (1984a) calculate abundances of C_2CN and HC_2CN that are within a factor of three of observed values. However, for the hydrocarbons C_3H_4 and C_4H, they calculate abundances that are low by at least two orders of magnitude.

In the realm of "chemical time dependent" treatments, Leung, Herbst, and Huebner (1984) have published a rather large model containing 200 chemical species and over 1800 gas phase reactions. The abundances of a variety of complex neutral molecules are calculated including HC_2CN, HC_4CN, CH_3OH, CH_3NH_2, CH_3CN, CH_3CHO, C_2H_5OH, and CH_3OCH_3. In general, these authors find that the abundances of complex molecules show an unusual time dependence not shown by most of the smaller species. Complex molecules tend to go through peak abundances at times between 10^5–10^6 yrs and then tend to diminish in abundance as steady-state is reached (Fig. 4). Agreement between observation and theory for these species is best at times well before steady-state is approached but after the peak abundances are achieved. The reason for this time dependence, which is far more severe than for most smaller species, is presumably the time dependence of the atomic carbon abundance in the model which is similar to that of the complex molecules. In this sense, it can be said that the chemical time dependent approach is in good agreement with the simpler "semi-detailed" steady-state method. In both models, atomic carbon appears to play a major role in complex molecule synthesis. The work of Leung, Herbst, and Huebner (1984) has been extended by Herbst and Leung (1986) to include the sensitivity of the calculated abundances to changes in temperature, cosmic abundance ratios, and photon flux. The inclusion of photons was undertaken because it is not at all clear that dense interstellar clouds are as opaque in the ultraviolet as is commonly believed. In general, Herbst and Leung (1986) did not find any major sensitivities of the calculated complex molecule abundances to small changes in the physical conditions.

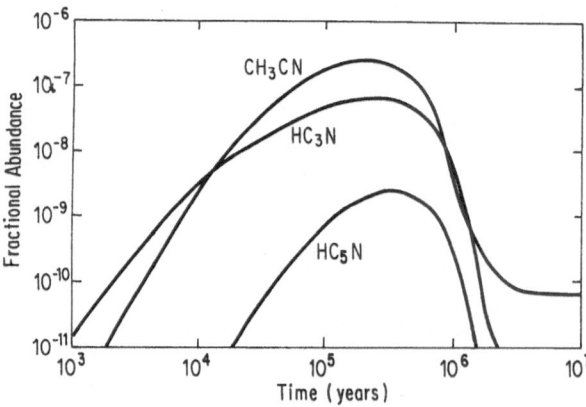

Fig. 4. The calculated time dependence of the fractional abundances of assorted complex molecules as obtained by Leung et al. (1984). The particular calculation shown utilizes a gas density of 2×10^4 cm^{-3}

What should we conclude from the time-dependent calculations? The most important conclusion is that gas phase models can produce a wide assortment of complex molecules in approximately the right abundances. Much more work remains to be done, however, in including molecules such as CH_3C_4H, CH_3C_5N, and higher order cyanoacetylenes into the models. It would appear that the "semi-detailed" approach offers the most realistic opportunity for inclusion of these and other complex molecules not yet observed, given its relatively small demands on computer space and time. Despite the apparent success of the time dependent model in obtaining large abundances of selected complex molecules, one should be cautious in interpreting the time dependence of the calculated complex molecule abundances. One cannot say that dense clouds are only say 10^6 years past the diffuse cloud stage based on the model results because the chemical time dependent approach is artificial in several ways. Firstly, the initial conditions utilized resemble a typical diffuse cloud except that the density is too large for a diffuse cloud. Secondly, the physical conditions of a cloud will change as the chemistry proceeds — in particular, the cloud will collapse on a time scale that is comparable to the time scale of the chemistry. These objections to the model can be overcome by putting physical collapse into the model. Several sets of investigators have already considered the hydrodynamics of cloud collapse along with simpler chemical networks (Gerola and Glassgold 1978; Tarafdar et al. 1985) and Herbst and Leung are preparing to consider the effect of cloud collapse on their reaction network. It is expected that if one starts the calculation out with a low density, representative of diffuse clouds, then the time scale for large molecule formation will lengthen. Even this degree of complexity may not be sufficient, however, because the problem of the interaction of the gas and dust remains; in particular, how do the molecules avoid adsorption on the dust grains? If one makes the standard assumption about grain densities and sizes (Sect. 5.3), grain adsorption times are much shorter than the time needed to reach chemical steady-state and similar to the time needed for maximum complex molecule abundances to be reached. That, we emphasize, does not prove that clouds are "young"; rather it proves that reality is more complex than current gas phase models.

4.1 Shock Calculations and Complex Molecules

The gas phase chemistry we have discussed up to now pertains to the ambient conditions in the interstellar medium. However, there is strong evidence that these ambient conditions are frequently changed by violent events known as interstellar shocks (see, e.g., Federman 1982; Elitzur and Watson 1980; Hartquist et al. 1980). An interstellar shock is a coherent disturbance passing through the cloud at speeds of perhaps 5–15 km sec^{-1} which results in an immediate large increase in temperature from below 100 K to at least several thousand K. The maximum post-shock temperature roughly depends on the square of the shock propagation velocity (Mitchell 1984 and references therein). An increase in density also occurs. Depending on the velocity of the shock, temperatures may become sufficiently high to destroy most molecules (Dalgarno and Roberge 1979). The high post-shock temperatures immediately begin to cool after the passage of the shock front and eventually return to normal in a time that depends on the gas density. The cooling is accompanied by an increase in gas density. For example, Mitchell (1984) has studied the effect of shocks on a dense cloud of

density 10^4 cm^{-3}. He finds that thirty years after the passage of the shock, the temperature has returned to 50 K and the density has increased to approximately 10^6 cm^{-3}.
Cooling occurs via inelastic collisions. However, the fact that spontaneous emission
rates of molecular vibrations and even rotations are shorter than the time scale for
cooling means that a complex non-equilibrium situation develops in which the vibrational temperature of most species is quite low, the rotational temperature is intermediate, and the translational temperature is highest. Unfortunately, most models
of the chemistry of interstellar shocks do not take this departure from equilibrium
into account and contain rate coefficients determined at thermal equilibrium in the
laboratory. Methods for determining rate coefficients under the non-thermal conditions present in the post-shock gas have been discussed by Herbst and Knudsen (1981)
and Herbst (1985 d).

The normal approach to the chemistry of the post shock gas is to add gas phase
reactions which, while prohibited from occurring under ambient conditions, can now
occur at the elevated temperatures present. These processes include exothermic
reactions with small activation energy barriers (e.g. neutral-neutral reactions) as
well as slightly endothermic reactions. For example, a neutral-neutral synthesis of
methane can occur (Mitchell 1984) via the following reactions in which the activation
energy in terms of temperature is included:

$$C + H_2 \rightarrow CH + H ; \quad \text{Activation Energy} = 14{,}100 \text{ K} \quad (4.1)$$

$$CH + H_2 \rightarrow CH_2 + H ; \quad \text{Activation Energy} = 8{,}800 \text{ K} \quad (4.2)$$

$$CH_2 + H_2 \rightarrow CH_3 + H ; \quad \text{Activation Energy} = 6{,}400 \text{ K} \quad (4.3)$$

$$CH_3 + H_2 \rightarrow CH_4 + H ; \quad \text{Activation Energy} = 5{,}500 \text{ K} . \quad (4.4)$$

Such a synthesis would be unthinkable under ambient conditions.

What effect do shocks have on the gas phase synthesis of complex interstellar
molecules? This question has been investigated at least for hydrocarbons through
six carbon atoms in complexity by Mitchell (1983, 1984). He has found that if a shock
passes through a dense cloud where much of the carbon is already in the form of carbon
monoxide, complex hydrocarbons are not formed in high abundance. However, if
a shock passes through a diffuse cloud, of density approximately 10^3 cm^{-3}, where
much of the cosmic abundance of carbon is in the form of C$^+$ and to a lesser extent C,
a different scenario is present. As the shock cools, the C$^+$ and C, which remain in
appreciable abundance for up to 10^5 yrs after the shock passage, react via many of the
reactions discussed above as well as others to produce a rich hydrocarbon chemistry.
The net effect is that large abundances of hydrocarbons build up as the cloud cools
and eventually reaches a gas density of 3×10^4 cm^{-3}. Do these results bear any relation to the results obtained from ambient gas phase models? In both types of calculations, hydrocarbon chemistry appears to require the presence of C$^+$ and/or C both
to synthesize one-carbon hydrocarbons such as methane and then, via insertion reactions, to produce more complex hydrocarbon species. Condensation reactions do
not appear to be sufficient.

5 Grains and Complex Molecules

In our discussion, we have considered only the gas phase formation and destruction of complex interstellar molecules. It is now time to consider how the presence of grains affects the abundances of these complex molecules. Let us briefly review what is meant by grains. Presumably these sub-micron sized particles contain a reasonably refractory core of silicates, amorphous carbon or graphite, and other materials, and a mantle of adsorbed species such as various ices and CO. A recent discussion on grain properties and observations is contained in the proceedings of the Hilo Workshop (Wolstencroft and Greenberg 1984). Background material can be found in Spitzer (1978). Grains are formed in high pressure regions such as cool stellar atmospheres and novae and ejected into the interstellar medium where they and associated gas condense to form clouds.

The presence of grains can result in several processes that affect complex molecules. First, synthetic reactions can occur on the surfaces of and inside grains to produce complex molecules. We shall discuss some views on how these processes might occur. There is some observational evidence for the presence of complex molecules on dust surfaces and we shall discuss the evidence. Secondly, grains act as sites for adsorption for species produced in the gas phase, and the evidence for complex molecules on grains does not tell us that they were formed on the grains. Indeed, there is a question as to how any molecular species other than hydrogen can ever leave the grains, independently of where and how they were synthesized. We shall therefore end this section by discussing some current views on how the gas phase is maintained in interstellar clouds. Thirdly, grains provide a shield behind which molecules can grow without being photodestroyed in a short time (perhaps as short as thirty years) by the interstellar ultra-violet radiation field.

5.1 Molecular Synthesis on and in Grains

Perhaps the dominant view of molecular formation on grains was discussed originally by Watson and Salpeter (1972a, b). In this view the grains are passive entities on which molecules adsorb, mainly via weak van der Waals forces, and are not catalytic agents in a Fisher-Tropsch sense. Watson and Salpeter (1972a, b) considered a cloud with an atomic gas as an initial condition and allowed the atoms to adsorb onto the cold grains present in the cloud. Once adsorbed on the surface, hydrogen atoms could migrate rather freely and associate with other hydrogen atoms as well as heavier atoms such as carbon, nitrogen, and oxygen which could migrate to less of an extent. Chemical bonds could be formed via the mechanism of the grain taking the energy of the bonds and storing it as thermal energy. In a sense, the grain would be acting as a third body in three body association reactions. After diatomic species such as CH, NH, and OH had formed, the process of H atom addition to reactive species could continue until saturated molecules such as methane, ammonia, and water were synthesized. More complex molecules could be formed by association reactions only if there were no activation energy involved in the reaction because the grains were not considered to be catalysts that could remove activation energy barriers. Getting the synthesized species off the grains presented a difficult problem and Watson and

Salpeter (1972 a, b) suggested ejection during the chemical reaction forming the species (some of the reaction exothermicity would be converted into ejection kinetic energy) or photoejection, in which a photon could provide the energy necessary to desorb the species. Unfortunately, there do not appear to be many photons, at least of the energetic variety in dense interstellar clouds, and Watson (1983) has recently stated that "except for molecular hydrogen, there is in my opinion still no convincing mechanism for returning molecules from grain surfaces back to the gas." In any event, in the model of Watson and Salpeter (1972 a, b), once back in the gas, the newly synthesized molecules would be destroyed via gas phase reactions.

The next major advance in grain synthesis was a series of papers by Allen and Robinson (1975, 1976, 1977) which are of major interest here because they contain "complex" organic molecules through eleven atoms in size. These authors suggested that some interstellar grains were sufficiently small that the energy of an exothermic chemical reaction could heat up a grain enough to release adsorbed species via thermal evaporation. This possibility has remained controversial. Another possibility had been suggested somewhat earlier by Aannestad (1973) — perhaps interstellar shock waves, discussed above, could drive adsorbed species off grains in the hot post-shock gas. The mechanism utilized by Allen and Robinson to produce complex organic molecules on grains was an extension of the Watson and Salpeter (1972 a, b) view that association reactions between atoms and radicals without activation energy could occur on passive grain surfaces. Unlike the Watson and Salpeter (1972 a, b) picture, Allen and Robinson did not consider the synthesized molecules to be depleted in the gas via gas phase reactions, but rather to be depleted only by readsorption onto the grains. Allen and Robinson (1977) presented calculated abundances in the gas phase for 372 molecules as functions of time and compared their results with observation. Perhaps a general comment about their work is that their grain model is necessarily, given our limited knowledge of the grain surface as well as grain chemistry, less selective in its predictions than are gas phase models. Still, if one assumes that there really is some mechanism capable of desorbing molecules from the surfaces of grains, the Allen and Robinson (1977) model does result in the production of complex molecules in the gas phase.

One difference between gas phase models and the grain model of Allen and Robinson (1977) is the predicted degree of saturation of the organic species. As we have seen, gas phase, ion-molecule models have difficulty in hydrogenating molecular ions with the result that highly non-saturated molecules are favored. This selectivity is not developed in the Allen and Robinson (1977) grain model. Although highly non-saturated species such as the cyanoacetylenes and alkynes are prominent in interstellar spectra, there are observational reasons for their prominence and it is unclear whether they truly dominate the interstellar gas.

A more recent model of grain chemistry in the same vein as the above authors has been undertaken by Tielens and Hagen (1982). In this model, the newly formed molecules on the grain surfaces remain on the grains, with the exception of molecular hydrogen, because of the lack of reasonable desorption mechanisms. Tielens and Hagen (1982) utilized the gas phase model of Prasad and Huntress (1980 a, b) to determine a set of gas phase abundances for adsorption onto the grains. They then utilized a subset of the Allen and Robinson (1977) reaction set, eliminating a large number of minor reactions and including some new reactions involving sulfur and oxygen atoms. Tielens and Hagen (1982) were interested in following the grain chemistry as a function of time to help explain some infra-red observations which are normally interpreted as caused by molecules on grains (see below).

In contrast to the rather dominant view of surface reactions on passive grains, Greenberg and collaborators (see, e.g., Greenberg 1984a, b) regard the grains as sites in which photochemistry can occur. To strengthen this position, this group has undertaken laboratory experiments on simulated interstellar grain photochemistry and has found that a mixture of complex molecules called "yellow stuff" can be synthesized from relatively simple precursors (CO, NH_3, H_2O, CH_4) and that a mechanism exists to get these species off the grains. This mechanism has been labeled "interstellar grain explosions" (d'Hendecourt et al. 1982); in the laboratory it is observed when the solid material that has been photolyzed with vacuum UV radiation is warmed to circa 27 K. Presumably, what is happening is a runaway series of reactions initiated by the free radicals resulting from the photolysis. In the interstellar medium, heating to 27 K could occur via grain-grain collisions (d'Hendecourt et al. 1982) or, as discussed in 5.3, via heavy nucleus cosmic ray bombardment. The major difficulty in this line of reasoning for dense clouds is that it requires a significant ultra-violet photon flux in regions normally thought of as relatively free of hard photons. Even if such a flux were present, its synthetic effect via grain photochemistry might not counterbalance its photodissociative effect in the gas. More work is needed to attempt to elucidate both the sources of radiation in dense clouds and the radiative transfer through these regions to determine the relative importance of this type of photochemistry. In addition, more work is needed to relate the simulations carried out by Greenberg's group to the interstellar medium where, even in unshielded regions, the radiation field is far less intense than in the laboratory.

5.2 Observations of Molecules on Grains

Although the discovery of interstellar molecules has been accomplished principally by radioastronomers searching for rotational transitions of gaseous species, there is some evidence for adsorbed molecules on grains as well. This evidence is in the form of broad spectral features seen in the infrared, both in absorption and emission. The emission comes from regions considerably warmer than the ambient interstellar medium such as star-forming areas or areas surrounding hot, newly-formed stars or is fluorescent in origin and the absorption is seen against such bright sources. (In this section, we will not discuss the infrared spectra of gaseous molecules seen in sources such as the stellar envelope IRC + 10216. Such observations yield information similar to radioastronomical studies with the exception that molecules that do not possess a permanent dipole moment such as CH_4 and C_2H_2 can be observed in the IR.) The broadness of the features makes for a certain ambiguity in their assignment and interpretation not present in the sharp gas phase spectra. Still, it is probable that features due to carbon monoxide, assorted ices, and silicates have been detected. A general discussion is to be found in the model paper by Tielens and Hagen (1982). What concerns us here is the evidence for complex organic molecules on grains. This evidence is particularly ambiguous.

The experiments of Greenberg and co-workers, discussed above, show that photo-processing of simulated laboratory grains can lead to a yellow organic residue (Greenberg 1984a, b). The infra-red spectra of these residues show a variable feature at 3.4 microns that is similar to a feature observed in the interstellar medium. Moore and Donn (1982) have utilized proton irradiation of low temperature ices of water, ammonia, and methane or propane to synthesize another complex organic residue which also shows the 3.4 micron feature. This wavelength coincides with the C—H stretching vibration and it is clear that a broad feature at this wavelength could indicate the presence of a variety of organic species. Indeed, Duley and Williams (1981) have suggested large polycyclic aromatic hydrocarbons, Hoyle and Wickramasinghe (1977) polysaccharides and Hoyle et al. (1982) biological species!

The feature at 3.4 microns is not the only unidentified interstellar infra-red spectral peak. In particular, there are bands at 6.2 and 7.7 microns which have recently been shown (Allamandola, Tielens, and Barker 1985) to bear a strong resemblance to the spectral features of automobile exhaust soot in this region. These authors believe the carriers of the interstellar features to be polycyclic aromatic hydrocarbons (Duley and Williams 1981) but feel these species to exist as individual molecules rather than in bulk grains. Note that the radius of a typical grain is expected to be around 1000 Å. An individual polycyclic aromatic hydrocarbon would be much smaller than this; Allamandola, Tielens, and Barker (1985) consider species of 35 Å or so. Presumably it is more appropriate to call such individual molecules a "gas" rather than "grains" since there is evidence that their excitation does not conform to normal bulk considerations. Even more recent work (Crawford et al. 1985) suggests that the polycyclic species are responsible for a series of diffuse bands in the optical region of the spectrum produced when starlight passes through diffuse interstellar clouds.

It is indeed physically possible that these polycyclic aromatic hydrocarbons could have been formed in a variety of stellar environments as the grains were being formed, and then ejected into the interstellar medium. Unlike smaller molecules, these species could be stable against photodissociation during the transit from source to cloud because, like grains, they can be considered as thermodynamic entities that convert light energy into heat. Calculations of their lifetimes in the unshielded interstellar radiation field would be most welcome. In addition, these molecules would be subjected to chemical reactions in the interstellar medium which would process them to some extent; this too should be investigated.

It does not seem likely, based on current interstellar cloud models of *in situ* synthesis, that molecules of such complexity could be synthesized in the interstellar environment, unless a photochemical mechanism on grains could be invoked. However, even such a mechanism is not normally so specific as to form these species as opposed to many other types of organic molecules.

5.3 Why Does a Gas Phase Exist?

The problem of the existence of a gas phase in dense interstellar clouds arises because the time scale for condensation onto grains appears to be shorter than the life time of these regions and, until recently, it has not been felt by many investigators that there was a viable mechanism for returning molecules heavier than H_2 back into the gas.

Let us first consider the time scale problem. The time t that will transpire before a molecule strikes a grain surface is given by the kinetic equation

$$t = 1/(v\sigma n_g) \tag{5.1}$$

where n_g (cm^{-3}) is the number density of grains, σ is the geometric cross section (cm^2) of the grains, and v (cm s^{-1}) is the speed of the molecule. Based on observations of clouds where some light penetrates the dust, it is known that the grains are approximately 0.1 microns in radius and contain about 1 % of the cloud mass. If it is assumed that the grains have a density of 3 gm/cm^3, representative of a core of silicate materials, then it is easily shown that

$$n_{H_2}/n_g = 3.8 \times 10^{11} \tag{5.2}$$

where n_{H_2} is the hydrogen gas density. If, in addition, it is assumed that v is 10^4 cm s^{-1}, a respresentative number for heavy species at interstellar cloud temperatures, then one can obtain from equations (5.1) and (5.2) that

$$t \text{ (yrs)} = 3 \times 10^9/n_{H_2} . \tag{5.3}$$

For a typical dense cloud with $n_{H_2} = 10^4$ cm^{-3}, $t = 3 \times 10^5$ yrs. Since the grains are quite cold, it is customarily assumed that hitting and sticking are one and the same for heavy species. The number we have calculated for t is a factor of 30 below the time needed to reach chemical steady state (see Section 4) and, indeed, is comparable to the time needed for peak complex molecule abundances to be achieved in the gas phase model of Leung, Herbst, and Huebner (1984). However, t is also much smaller than customarily assumed cloud lifetimes, based on star formation rates, of 10^7 yrs (see, e.g., Leger, Jura, and Omont 1985). If this latter number is correct, then a gas phase can exist if and only if at least one of the following criteria is met:

a) the clouds being observed are preferentially young clouds,
b) heavy molecules have a low sticking probability on grains despite the low temperature
c) there is an efficient desorption mechanism.

Possibility a) seems remote based on statistical grounds and possibility b) has not received much attention. It might be possible that H$_2$ is retained on grain surfaces long enough before thermal evaporation to form a monolayer which then hinders adsorption by heavier species which do not stick well to molecular hydrogen. However, such a scenario requires a significant thermal evaporation time for molecular hydrogen after it is formed on grains or sticks to the grain surface. More probable is possibility c) although most investigators have had difficulty in ascertaining what the possible mechanism for grain desorption might be. Possibilities we have already discussed include grain explosions, interstellar shocks, and photodesorption.

Quite recently, Leger, Jura, and Omont (1985) have carefully investigated desorption mechanisms from interstellar grains. These authors have concentrated on grain heating via X-rays and cosmic rays followed by classical thermal evaporation or grain

165

explosion. The most important contribution in dark regions was found to be caused by heavy cosmic ray particles because, despite the relative rarity of these species, they deposit more energy into the grains. The authors distinguish between CO-rich mantles and mantles of refractory ices (H_2O, NH_3, H_2CO) with higher heat capacities. For the latter mantles, they find very little desorption whereas for CO-rich mantles, they find that desorption can occur to a limited extent and effectively increase t in equation (81) to t′ where

$$t' \text{ (yrs)} = 10^{11}/n_{H_2} . \tag{5.4}$$

Thus, for a gas density of 10^4 cm^{-3}, t′ is approximately 10^7 yrs, which is comparable to the cloud lifetime. In this view, condensation ultimately overwhelms desorption, but at a time scale too long for importance in cloud models. Note that t′ refers to all heavy molecules as long as they reside in a CO-rich mantle. Of course, photodesorption via UV photons from internal cloud sources (e.g., stars formed inside clouds), turbulent grain-grain collisions, and interstellar shocks can also aid in increasing the rate of desorption and effectively increasing t′ or even achieving a steady-state in which a certain fraction of heavy molecules are "always" in the gas phase.

The work of Leger, Jura, and Omont (1985) clearly differs from earlier, more pessimistic conclusions concerning desorption. If correct, its implications are that a significant fraction of molecules will continue to exist in the gas phase during the cloud lifetime, and that some molecules formed on grain surfaces can desorb into the gas. Thus, the existing controversy between the proponents of gas phase and grain surface mechanisms for molecule synthesis remains unresolved by this work.

6 How Large can Interstellar Organic Molecules be?

In this last section, we speculate on two related questions of great interest: what if any is the limit of the complexity attainable by interstellar molecules and how can increasingly complex molecules be detected? Our answer to the first question derives in part from the mechanism assumed to form the interstellar organic species. Let us first consider *in situ* syntheses. If one looks at the results of those models, both gas phase and grain, that contain complex molecules (viz., Leung, Herbst, and Huebner 1984; Allen and Robinson 1977) one sees that as organic molecules become more complex their abundances decrease significantly. This effect seems reasonable if not obvious; after all there is a finite abundance of the element carbon and if the abundance of organic molecules increased with complexity ad infinitum, an obvious contradiction would result. An observational example of this trend is seen for the cyanoacetylenes observed in the TMC-1 interstellar cloud. Here, according to Broten et al. (1978), as one proceeds to the next most complicated member of the series the abundance drops off by a fairly constant factor of four. As of the present, the largest observed member of this series in TMC-1 is $HC_{10}CN$ (Bell and Matthews 1985). If one extrapolates this factor four to larger and larger cyanoacetylenes, one sees that the abundances of these species become quite small indeed and it would appear that these molecules will be increasingly difficult to detect.

What are the prospects for observing these and other more complex species via radioastronomical methods? As molecules become larger, their rotational densities of states become greater and populations in specific rotational states become smaller. Thus, intensities of individual rotational transitions become lower even if abundances remain the same (Herbst 1985e) and increasingly large species will be increasingly difficult to detect via rotational spectra. This objection pertains to high resolution visible, ultra-violet, and infra-red spectra as well as for such spectra, which involve individual rotational states. Still, as telescope receivers improve, the prospects for observing more complex molecules will also improve. The process will probably be an incremental rather than a dramatic one as slightly more complex species than those previously identified are continually discovered. Diffuse bands in the infra-red and visible, discussed above, probably represent a more realistic hope of observing very complex species, but these broad features are far more difficult to identify. Infra-red detection of bonds and functional groups rather than whole molecular structures would appear to be more likely as can be seen from the controversy concerning the 3.4 micron feature.

One model calculation in which increasing complexity does not lead to lower abundances in a certain sense is that of Herbst and Leung (1985) in which, following a suggestion of Langer et al. (1984), the carbon elemental abundance is set to a greater value than that of oxygen in the gas phase. There are stars which show a greater abundance of carbon than of oxygen, called carbon stars, but they appear to be heavily outnumbered by oxygen-rich stars. Therefore, if the carbon abundance in the gas phase of clouds is to exceed that of oxygen, the principal mechanism would have to be selective depletion of oxygen onto the grains. In any event, this model shows that for hydrocarbons ranging up to four carbon atoms in size, the overall abundance of all hydrocarbons with a given number of carbon atoms does not decrease as the number of carbon atoms increases. Boldly extrapolating this feature of the model to larger hydrocarbons results in the prediction that there are significant amounts of very complex hydrocarbons in interstellar clouds. Of course, even if this speculative argument were true and even if the high carbon abundance did pertain to some regions of interstellar space, one must remember that as the number of carbon atoms increases, the number of isomers of each hydrocarbon increases dramatically. Consequently, unless the chemistry is a most selective one, individual complex molecules will still be difficult to observe. And, of course, there is still the observational problem caused by the large number of rotational states in complex molecules.

If very large interstellar molecules exist in appreciable abundance and are formed in the clouds, it would appear most likely based on these limited results that they are formed in regions where the cosmic abundance of carbon exceeds that of oxygen in the gas phase. The gas phase of a cloud in which oxygen is less abundant than carbon resembles the reducing atmosphere postulated to exist by some investigators on the early earth where organic species could have been formed via a Urey-Miller type synthesis. Of course, there is great debate over the constituents of the early earth atmosphere, with current wisdom holding that the atmosphere was not as reducing as once thought and that significant abundances of species such as CO and CO_2 might have existed. Still, in both the interstellar and atmospheric contexts, it would appear that molecular complexity is best achieved in a high carbon, reducing atmosphere.

Let us return to the question of external formation processes for complex molecules.

After all, astronomers agree that dust particles are formed in high pressure, stellar atmospheric regions. Why not individual complex molecules as well? The argument up to now has been that complex molecules could not survive the journey to an interstellar cloud where the dust particles would protect them against hard ultra-violet radiation. In addition, there has been the observational argument that complex molecules are not detected in more than several stellar-type sources. Both of these negative arguments are open to some challenge. As molecules become more complex and have more modes of energy disposal, it becomes more possible for them to absorb hard ultra-violet radiation and randomize the energy over many modes until they can reradiate the energy rather than photodissociate. In this sense, they can be considered as solid, bulk materials. In addition, gas phase organic molecules through $HC_{10}CN$ in complexity have been observed in the carbon-rich circumstellar source IRC + 10216 (Lafont, Lucas, and Omont 1982) where the presumed mechanism for their formation involves thermodynamic equilibrium modified by photochemistry. Broad infra-red features attributable to complex molecules have also been seen emanating from high density, high temperature regions (see the discussion in Sect. 5.2). It seems reasonable to suggest that more effort will be made in the future to understand whether or not very large interstellar molecules can indeed be formed in stellar and quasi-stellar sources and ejected into clouds.

Given the uncertainties in our understanding of interstellar and circumstellar chemistry, it is not unreasonable to speculate that there may well be a large number of complex molecules smaller than individual grains that exist in both circumstellar and interstellar sources. However, much more work remains to be done on characterizing the spectra of such complex species in all regions of the spectrum and of determining how and where they can be synthesized. The detection of increasingly complex molecules in other regions of the cosmos is a fascinating intellectual exercise and has implications for the study of how life arose on earth and possibly other sources. After all, interstellar clouds are known to be the birthplace of stars and at least one star is known to contain a planetary system. Interstellar organic molecules, if suitably protected, can emerge unscathed as a star is formed, perhaps embedded in comets. The relation between interstellar organic molecules and those on earth may be far stronger than is currently realized. In any event, the search for more and larger organic molecules in space will remain an exciting if difficult quest.

Note Added in Proof

New Molecules

At a conference on Astrochemistry held during December 1985 at Goa, India, we learned of probable identifications of two new molecular ions in interstellar clouds. Woods (1985) discussed the identification of several lines of SO^+ in Orion, based on the laboratory work of his group. Ziurys and Turner (1985) discussed the observation of three rotational transition frequencies of the linear ion $HCNH^+$ in the source Sgr B2, based on the laboratory work of Bogey et al. (1985). The large $HCNH^+/HCN$

abundance ratio, derivable from the spectra, supports the idea of Adams, Smith, and Clary (1985) that ion-polar reactions such as

$$HCN + H_3^+ \rightarrow HCNH^+ + H_2$$

are especially rapid at interstellar cloud temperatures.

In addition to the probable detection of these two molecular ions, a weak line at 234.9 GHz in Orion has been tentatively identified by Sutton et al. (1985) as the $J = 5 \rightarrow 4$ transition frequency of PN, based on laboratory data from Wyse et al. (1972).

7 Acknowledgements

G. W. wishes to thank the Deutsche Forschungsgemeinschaft for financial support via grant SFB-301. E.H. acknowledges support of his theoretical program by the National Science Foundation (U.S.) through grant AST-8312270.

8 References

Aannestad, P.: Astrophys. J. Suppl. *25*, 223 (1973)

Adams, N. G., Smith, D.: Chem. Phys. Lett. *79*, 563 (1981)

Adams, N. G., Smith, D., Clary, D. C.: Astrophys. J. (Letters) *296*, L31 (1985)

Allamandola, L. J., Tielens, A. G. G. M., Barker, J. R.: Astrophys. J. (Letters) *290*, L25 (1985)

Allen, M., Robinson, G. W.: Astrophys. J. *195*, 81 (1975)

Allen, M., Robinson, G. W.: ibid. *207*, 745 (1976)

Allen, M., Robinson, G. W.: ibid. *212*, 396 (1977)

Anicich, V. G., Huntress, W. T., Jr.: Jet Propulsion Laboratory preprint (1984)

Barlow, S. E., Dunn, G. H., Schauer, M.: Phys. Rev. Letters *52*, 902 (1984)

Bass, L. M., Cates, R. D., Jarrold, M. F., Kirchner, N. J., Bowers, M. T.: J. Amer. Chem. Soc. *105*, 7024 (1983)

Bass, L. M., Chesnavich, W. J., Bowers, M. T.: ibid. *101*, 5493 (1979)

Bates, D. R.: J. Phys. B: Atom Molec. Phys. *12*, 4135 (1979)

Bates, D. R.: Astrophys. J. *270*, 564 (1983)

Bell, M. B., Matthews, H. E.: Astrophys. J. (Letters) *291*, L63 (1985)

Bester, M., Yamada, K., Winnewisser, G., Joentgen, W., Altenbach, H.-J., Vogel, E.: Astron. Astrophys. *137*, L20 (1984)

Blake, G. A.: Ph. D. Thesis (Calif. Inst. Tech.) (1985)

Blake, G. A., Helminger, P., Herbst, E., De Lucia, F. C.: Astrophys. J. (Letters) *264*, L69 (1983)

Blake, G. A., Keene, J., Phillips, T. G.: Astrophys. J., *295*, 501 (1985)

Bogey, M., Demuynck, C., Destombes, J. L.: Astron. Astrophys. *138*, L11 (1984)

Bogey, M., Demuynck, C., Denis, M., Destombes, J. L., Lemoine, B.: Astron. Astrophys. *137*, L15 (1984)

Bogey, M., Destombes, J. L.: Astron. Astrophys. (Letters) (in press) (1986)

Bohme, D. K.: private communication (1985)

Bohme, D. K., Raksit, A. B.: M.N.R.A.S. *213*, 717 (1985)

Broten, N. W., Oka, T., Avery, L. W., MacLeod, J. M.: Astrophys. J. (Letters) *223*, L105 (1978)

Brown, R. D., Eastwood, F. W., Elmes, P. S., Godfrey, P. D.: J. Amer. Chem. Soc. *105*, 6496 (1983)

Brown, R. D., Godfrey, P. D., Elmes, P. S., Rodler, M., Tack, L. M.: ibid. *107*, 4112 (1985)

Cheung, A. C., Rank, D. M., Townes, C. H., Thornton, D. D., Welch, W. J.: Phys. Ref. Letters *21*, 1701 (1968)

Crawford, M. K., Tielens, A. G. G. M., Allamandola, L. J.: Astrophys. J. (Letters) *293*, L45 (1985)
Cummins, S. E., Linke, R. A., Thaddeus, P.: submitted to Astrophys. J. Suppl. (1985)
Dalgarno, A., Roberge, W. G.: Astrophys. J. (Letters) *233*, L25 (1979)
d'Hendecourt, L. B., Allamandola, L. J., Baas, F., Greenberg, J. M.: Astron. Astrophys. *109*, L12 (1982)
De Frees, D. J., McLean, A. D., Herbst, E.: Astrophys. J. *279*, 323 (1984)
De Frees, D. J., McLean, A. D., Herbst, E.: ibid. *293*, 236 (1985)
De Lucia, F. C., Herbst, E., Plummer, G., Blake, G. A.: J. Chem. Phys. *78*, 2312 (1983)
De Lucia, F., Helminger, P., Gordy, W.: Phys. Rev. *A 3*, 1849 (1971)
Duley, W. W., Williams, D. A.: M.N.R.A.S. *196*, 269 (1981)
Duley, W. W., Williams, D. A.: Interstellar Chemistry, Academic Press, London 1984
Elitzur, M., Watson, W. D.: Astrophys. J. *236*, 172 (1980)
Federer, W., Villinger, H., Howorka, F., Lindinger, W., Tosi, P., Bassi, D., Ferguson, E. E.: Phys. Rev. Letters *52*, 2084 (1984)
Federman, S. R.: Astrophys. J. *253*, 601 (1982)
Gerola, H., Glassgold, A. E.: Astrophys. J. Suppl. *37*, 1 (1978)
Gordy, W., Cook, R. L.: Microwave Molecular Spectra, J. Wiley, New York 1984
Gottlieb, C. A., Vrtilek, J. M., Gottlieb, E. W., Thaddeus, P., Hjalmarson, A.: Astrophys. J. (Letters) *294*, L55 (1985)
Graedel, T. E., Langer, W. D., Frerking, M. A.: Astrophys. J. Suppl. *48*, 321 (1982)
Green, S., Herbst, E.: Astrophys. J. *229*, 121 (1979)
Greenberg, J. M.: Scientific American *250*, 124 (1984a)
Greenberg, J. M.: Proceedings of Workshop on Laboratory and Observational Infrared Spectra of Interstellar Dust, Wolstencraft and Greenberg, eds., Occasional Reports of the Royal Observatory, Edinburgh 12, *82* (1984b)
Gudeman, C. S., Haese, N. N., Piltch, N. D., Woods, R. C.: Astrophys. J. (Letters) *246*, L47 (1981)
Gudeman, C. S., Woods, R. C.: Phys. Rev. Letters *48*, 1344 (1982)
Hartman, J.: Astrophys. J. *21*, 389 (1905)
Hartquist, T. W., Oppenheimer, M., Dalgarno, A.: ibid. *236*, 182 (1980)
Herbst, E.: ibid. *205*, 94 (1976)
Herbst, E.: ibid. *222*, 508 (1978)
Herbst, E.: ibid *237*, 462 (1980a)
Herbst, E.: ibid. *241*, 197 (1980b)
Herbst, E.: J. Chem. Phys. *75*, 4412 (1981)
Herbst, E.: Chem. Phys. *68*, 323 (1982a)
Herbst, E.: Astrophys. J. *252*, 810 (1982b)
Herbst, E.: Astrophys. J. Suppl. *53*, 41 (1983)
Herbst, E.: Astrophys. J. *291*, 226 (1985a)
Herbst, E.: J. Chem. Phys. *82*, 4017 (1985b)
Herbst, E.: Astrophys. J. *292*, 484 (1985c)
Herbst, E.: Astron. Astrophys. *153*, 151 (1985d)
Herbst, E.: Origins of Life in press (1986)
Herbst, E., Adams, N. G., Smith, D.: Astrophys. J. *269*, 329 (1983)
Herbst, E., Adams, N. G., Smith, D.: ibid. *285*, 618 (1984)
Herbst, E., Klemperer, W.: ibid. *185*, 505 (1973)
Herbst, E., Klemperer, W.: Physics Today *29*, 32 (1976)
Herbst, E., Knudson, S.: Astrophys. J. *245*, 529 (1981)
Herbst, E., Leung, C. M.: M.N.R.A.S. in press (1986)
Herbst, E., Smith, D., Adams, N. G.: Astron. Astrophys. *138*, L13 (1984)
Hollis, J. M., Snyder, L. E., Lovas, F. J., Ulich, B. L.: Astrophys. J. *241*, 158 (1980)
Hollis, J. M., Snyder, L. E., Blake, D. H., Lovas, F. J., Suenram, R. D., Ulich, B. L.: ibid. *251*, 541 (1982)
Hoyle, F., Wickramasinghe, N. C.: Nature *268*, 610 (1977)
Hoyle, F., Wickramasinghe, N. C., Al-Mufti, S., Olavesen, A. H.: Astrophys. Space Sci. *81*, 489 (1982)
Huntress, W. T., Jr.: Astrophys. J. Suppl. *33*, 495 (1977)
Huntress, W. T., Jr.: Mitchell, G. F.: Astrophys. J. *231*, 456 (1979)
Irvine, W. M., Schloerb, F. P.: ibid. *282*, 516 (1984)

Irvine, W. M., Schloerb, F. P., Hjalmarson, A., Herbst, E.: Protostars and Planets II, p 579, Univ. of Arizona, 1985

Jaffe, D. T., Harris, A. I., Silber, M., Genzel, R., Betz, A. L.: Astrophys. J. (Letters) 290, L59 (1985)

Jarrold, M. F., Bowers, M. T., De Frees, D. J., McLean, A. D., Herbst, E.: in preparation (1985)

Johannson, L. E. B., Andersson, C., Ellder, J., Friberg, P., Hjalmarson, A., Höglund, B., Irvine, W. M., Olofsson, H., Rydbeck, G.: Astron. Astrophys. 130, 227 (1984)

Keene, J., Blake, G. A., Huggins, P. J., Phillips, T. G., Beichman, C. A.: Astrophys. J. 299, 967 (1985)

Klebsch, W., Bester, M., Yamada, K., Winnewisser, G., Joentgen, W., Altenbach, H.-J., Vogel, E.: Astron. Astrophys. 152, L12 (1985)

Knight, J. S., Freeman, C. G., McEwan, M. J., Adams, N. G., Smith, D.: submitted to Int. J. Mass Spec. Ion Phys. (1985)

Lafont, S., Lucas, R., Omont, A.: Astron. Astrophys. 106, 201 (1982)

Langer, W. D., Graedel, T. E., Frerking, M. A., Armentrout, P. B.: Astrophys. J. 277, 581 (1984)

Leger, A., Jura, M., Omont, A.: Astron. Astrophys. 144, 147 (1985)

Leung, C. M., Herbst, E., Huebner, W. F.: Astrophys. J. Suppl. 56, 231 (1984)

Luine, J. A., Dunn, G. H.: Astrophys. J. (Letters) 299, L67 (1985)

Matthews, H. E., Irvine, W. M., Friberg, P., Brown, R. D., Godfrey, P. D.: Nature 310, 125 (1984)

Marquette, J. B., Rowe, B. R., Dupeyrat, G., Roueff, E.: Astron. Astrophys. 147, 115 (1985)

Michalopoulous, D. L., Geusic, M. E., Langridge-Smith, P. R. R., Smalley, R. E.: J. Chem. Phys. 80, 3556 (1984)

Millar, T. J., Freeman, A.: M.N.R.A.S. 207, 405 (1984a)

Millar, T. J., Freeman, A.: ibid. 207, 425 (1984b)

Mitchell, G. F.: ibid. 205, 765 (1983)

Mitchell, G. F.: Astrophys. J. Suppl. 54, 81 (1984)

Mitchell, G. F., Ginsburg, J. L., Kuntz, P. J.: ibid. 38, 39 (1978)

Mitchell, G. F., Huntress, W. T., Jr.: Nature 278, 722 (1979)

Mitchell, G. F., Huntress, W. T., Prasad, S. S.: Astrophys. J. 233, 102 (1979)

Moore, M. H., Donn, B.: Astrophys. J. (Letters) 257, L47 (1982)

Morris, M., Zuckermann, B., Palmer, P., Turner, B. E.: Astrophys. J. 186, 501 (1973)

Mul, P. M., McGowan, J. Wm.: ibid. 237, 749 (1980)

Phillips, T. G., Huggins, P. J.: ibid. 251, 533 (1971)

Phillips, T. G., Blake, G. A., Keene, J., Woods, R. C., Churchwell, E.: Astrophys. J. (Letters) 294, L45 (1985)

Plummer, G. M., Herbst, E., De Lucia, F. C.: J. Chem. Phys. 83, 1428 (1985)

Prasad, S. S., Huntress, W. T., Jr.: Astrophys. J. Suppl. 43, 1 (1980a)

Prasad, S. S., Huntress, W. T., Jr.: Astrophys. J. 239, 151 (1980b)

Rowe, B. R., Dupeyrat, G., Marquette, J. B., Smith, D., Adams, N. G., Ferguson, E. E.: J. Chem. Phys. 80, 241 (1984)

Rosenstock, H. M., Draxl, K., Steiner, B. W., Herron, J. T.: Energetics of Gaseous Ions (Suppl. to the J. Phys. Chem. Ref. data, Vol. 6, 1977)

Schiff, H. I., Bohme, D. K.: Astrophys. J. 232, 740 (1979)

Schwahn, G., Schieder, R., Bester, M., Winnewisser, G.: J. Mol. Spectrosc. (in press) (1985)

Smith, D., Adams, N. G.: Astrophys. J. (Letters) 220, L87 (1978)

Smith, D., Adams, N. G.: Int. Rev. Phys. Chem. 1, 271 (1981)

Smith, D., Adams, N. G., Alge, E., Herbst, E.: Astrophys. J. 272, 365 (1983)

Snyder, L. E., Wilson, T. L., Henkel, C., Jewell, R., Walmsley, C. M.: BAAS 16, 959 (1984)

Spitzer, L., Jr.: Physical Processes in the Interstellar Medium, New York: Wiley-Interscience 1978

Sutton, E. C., Blake, G. A., Masson, C. R., Phillips, T. G.: Astrophys. J. Suppl. 58, 341 (1985); ibid 60, 357 (1986)

Suzuki, H.: Astrophys. J. 272, 579 (1983)

Tang, B. T., Inokuchi, H., Saito, S., Yamada, Ch., Hirota, E.: Chem. Phys. Lett. 116, 83 (1985)

Tarafdar, S. P., Prasad, S. S., Huntress, W. T., Jr., Villere, K. R., Black, D. C.: Astrophys. J. 289, 220 (1985)

Thaddeus, P., Gottlieb, C. A., Hjalmarson, A., Johansson, L. E. B., Irvine, W. M., Friberg, P., Linke, R. A.: Astrophys. J. (Letters) 294, L49 (1985)

Thaddeus, P., Vrtilek, J. M., Gottlieb, C. A.: Astrophys. J. 299, L63 (1985a)

Thaddeus, P., Guélin, M., Linke, R. A.: Astrophys. J. 246, L41 (1981)

Thaddeus, P., Cummins, S. E., Linke, R. A.: ibid. *283*, L45 (1984)

Thorne, L. R., Anicich, V. G., Prasad, S. S., Huntress, Jr. W. T.: ibid. *280*, 139 (1984)

Tielens, A. G. G. M., Hagen, W.: Astron. Astrophys. *114*, 245 (1982)

Toelle, F., Ungerechts, H., Walmsley, C. M., Winnewisser, G., Churchwell, E.: ibid. *95*, 143 (1981)

Vrtilek, J. M., Thaddeus, P., Gottlieb, C. A.: BAAS. *17*, 568 (1985)

Walmsley, C. M., Winnewisser, G., Toelle, F.: Astron. Astrophys. *81*, 245 (1980)

Walmsley, C. M., Jewell, P. R., Snyder, L. E., Winnewisser, G.: ibid. *134*, L11 (1984)

Warner, H. E., Connor, W. T., Petrmichl, R. H., Woods, R. C.: J. Chem. Phys. *81*, 2514 (1984)

Watson, W. D.: Univ. of Illinois preprint ILL-(AST)-83-26 (1983)

Watson, W. D., Salpeter, E. E.: Astrophy. J. *174*, 321 (1972a)

Watson, W. D., Salpeter, E. E.: ibid. *175*, 659 (1972b)

Weinreb, S., Barrett, A. H., Meeks, M. L., Henry, J. C.: Nature *200*, 829 (1963)

Williams, D. A.: Astrophysical Letters *10*, 17 (1972)

Winnewisser, G.: Topics in Current Chemistry *99*, 39 (1981)

Winnewisser, G., Mezger, P. G., Breuer, H. D.: ibid. *44*, 1 (1974)

Winnewisser, G., Churchwell, E., Walmsley, C. M.: Modern Aspects of Microwave Spectroscopy (Chantry, G. W. ed.) Academic Press 1979

Winnewisser, G., Toelle, F., Ungerechts, H., Walmsley, C. M.: Proceedings IAU Symp. No. *87*, 59 (1980)

Winnewisser, G., Aliev, M. R., Yamada, K.: Proceed. ESA Workshop, ESA Sp-*189*, 23 (1982)

Winnewisser, G.: Winnewisser, M., Christiansen, J. J.: Astron. Astrophys. *109*, 141 (1982)

Winnewisser, G., Walmsley, C. M.: Astrophys. Spa. Science *65*, 83 (1979)

Wolstencraft, R. D., Greenberg, J. M., eds.: Proceedings of Workshop on Laboratory and Observational Infrared Spectra of Interstellar Dust, Occasional Reports of the Royal Observatory, Edinburgh *12*, (1984)

Woods, R. C., Gudeman, C. S., Dickman, R. L., Goldsmith, P. F., Huguenin, G. R., Irvine, W. M., Hjalmarson, A., Nyman, L.-A., Olofsson, H.: Astrophys. J. *270*, 583 (1983)

Woods, R. C.: Proceed. of IAU Symp. No. 120 on Astrochemistry (Reidel 1986 to be published)

Wyse, F. C., Manson, E. L., Gordy, W.: J. Chem. Phys. *57*, 1106 (1972)

Ziurys, L. M., Turner, B. E.: Astrophys. J. (Letters) *292*, L25 (1985)

Ziurys, L. M., Turner, B. E.: Proceed. of IAU Symp. No. 120 on Astrochemistry (Reidel 1986 to be published)

Author Index Volumes 101–139

174

Portmann, P., see Badertscher, M.: *136*, 17–80 (1986).
Pressman, B. C., see Painter, R.: *101*, 84–110 (1982).
Pretsch, E., see Badertscher, M.: *136*, 17–80 (1986).
Prinsen, W. J. C., see Laarhoven, W. H.: *125*, 63–129 (1984).

Rabenau, A., see Kniep, R.: *111*, 145–192 (1983).
Rauch, P., see Káš, J.: *112*, 163–230 (1983).
Raymond, K. N., Müller, G., and Matzanke, B. F.: Complexation of Iron by Siderophores A Review of Their Solution and Structural Chemistry and Biological Function. *123*, 49–102 (1984).
Recktenwald, O., see Veith, M.: *104*, 1–55 (1982).
Reetz, M. T.: Organotitanium Reagents in Organic Synthesis. A Simple Means to Adjust Reactivity and Selectivity of Carbanions. *106*, 1–53 (1982).
Reisse, J.: see Mullie, F., *139*, 83–117 (1986).
Rolla, R., see Montanari, F.: *101*, 111–145 (1982).
Rossa, L., Vögtle, F.: Synthesis of Medio- and Macrocyclic Compounds by High Dilution Principle Techniques. *113*, 1–86 (1983).
Rubin, M. B.: Recent Photochemistry of α-Diketones. *129*, 1–56 (1985).
Rüchardt, Ch., and Beckhaus, H.-D.: Steric and Electronic Substituent Effects on the Carbon-Carbon Bond. *130*, 1–22 (1985).
Rzaev, Z. M. O.: Coordination Effects in Formation and Cross-Linking Reactions of Organotin Macromolecules. *104*, 107–136 (1982).

Saenger, W., see Hilgenfeld, R.: *101*, 3–82 (1982).
Saller, H.: see Gasteiger, J., *137*, 19–73 (1986).
Sandorfy, C.: Vibrational Spectra of Hydrogen Bonded Systems in the Gas Phase. *120*, 41–84 (1984).
Schlögl, K.: Planar Chiral Molecural Structures. *125*, 27–62 (1984).
Schmeer, G., see Barthel, J.: *111*, 33–144 (1983).
Schmidt, G.: Recent Developments in the Field of Biologically Active Peptides. *136*, 109–159 (1986).
Schmidtchen, F. P.: Molecular Catalysis by Polyammonium Receptors. *132*, 101–133 (1986).
Schöllkopf, U.: Enantioselective Synthesis of Nonproteinogenic Amino Acids. *109*, 65–84 (1983).
Schuster, P., see Beyer, A., see *120*, 1–40 (1984).
Schwochau, K.: Extraction of Metals from Sea Water. *124*, 91–133 (1984).
Shugar, D., see Czochralska, B.: *130*, 133–181 (1985).
Selig, H., and Holloway, J. H.: Cationic and Anionic Complexes of the Noble Gases. *124*, 33–90 (1984).
Séquaris, J.-M., see Koglin, E.: *134*, 1–57 (1986).
Shibata, M.: Modern Syntheses of Cobalt(III) Complexes. *110*, 1–120 (1983).
Shinkai, S., and Manabe, O.: Photocontrol of Ion Extraction and Ion Transport by Photo-functional Crown Ethers. *121*, 67–104 (1984).
Shubin, V. G. Contemporary Problemsn Carbonium Ion Chemistry II. *116/117*, 267–341 (1984).
Siegel, H.: Lithium Halocarbenoids Carbanions of High Synthetic Versatility. *106*, 55–78 (1982).
Sinta, R., see Smid, J.: *121*, 105–156 (1984).
Smid, J., and Sinta, R.: Macroheterocyclic Ligands on Polymers. *121*, 105–156 (1984).
Soos, Z. G., see Keller, H. J.: *127*, 169–216 (1985).
Steudel, R.: Homocyclic Sulfur Molecules. *102*, 149–176 (1982).
Steudel, R., and Laitinen, R.: Cyclic Selenium Sulfides. *102*, 177–197 (1982).
Suzuki, A.: Some Aspects of Organic Synthesis Using Organoboranes. *112*, 67–115 (1983).
Suzuki, A., and Dhillon, R. S.: Selective Hydroboration and Synthetic Utility of Organoboranes thus Obtained. *130*, 23–88 (1985).
Szele, J., Zollinger, H.: Azo Coupling Reactions Structures and Mechanisms. *112*, 1–66 (1983).

Tabushi, I., Yamamura, K.: Water Soluble Cyclophanes as Hosts and Catalysts. *113*, 145–182 (1983).
Takagi, M., and Ueno, K.: Crown Compounds as Alkali and Alkaline Earth Metal Ion Selective Chromogenic Reagents. *121*, 39–65 (1984).
Tagaki, W., and Ogino, K.: Micellar Models of Zinc Enzymes. *128*, 143–174 (1985).
Takeda, Y.: The Solvent Extraction of Metal Ions by Grown Compounds. *121*, 1–38 (1984).